Physics with Example

for high school and college

By Şerife Sarıca

Copyright © 2014 by Şerife Sarıca

All rights reserved. No part of this publication may be reproduced, distributed, or transmitted in any form or by any means, including photocopying, recording, or other electronic or mechanical methods, without the prior written permission of the publisher, except in the case of brief quotations embodied in critical reviews and certain other noncommercial uses permitted by copyright law. For permission requests, write to the publisher, addressed "Attention: Permissions Coordinator," at the address below.

Melih SARICA
25883 N Park Ave
Unit A993106
Elkhart, Indiana, 46514, USA
Tel: 574-742-8531

I dedicate this book to my family including my husband who gave me the idea of writing and endless support, my cats (lucky and little guy) who make me happy and finally my dear little son who delayed my work a year with his mischief.

PREFACE

I am a physics teacher, before writing this book; I built two websites that include high school and college physics and chemistry subjects. While preparing these websites and this book, I focused on giving summary of the topic, highlight important points and solve more examples. Throughout my study life I saw many books that give lots of information and in my humble opinion in this case students may miss important points. Most of the time, instead of dealing with information overload, students prefer summary information and use cases; my websites' usage statistics support this theory too.

In this book, thirteen main physics subjects (vectors, mechanics, energy, work and power, impulse momentum, rotational motion, optics, properties of matter, heat temperature and thermal expansion, electrostatics, electric current, magnetism and waves) and their subtopics are explained. Moreover, extra examples are given at the end of each subject with their solutions, especially targeting high school and college students.

I hope this book will be helpful for you. Enjoy physics with the help of *physics with examples*!

PROPERTIES OF VECTORS	8
ADDITION OF VECTORS	9
MULTIPLYING A VECTOR WITH A SCALAR	10
COMPONENTS OF VECTORS	11
MORE PROBLEMS RELATED TO VECTORS	13
DISTANCE AND DISPLACEMENT	18
SPEED	19
VELOCITY	20
AVERAGE SPEED AND INSTANTANEOUS SPEED	21
AVERAGE VELOCITY AND INSTANTANEOUS VELOCITY	21
ACCELERATION	22
DESCRIBING MOTION WITH GRAPHS	25
FREE FALL	32
RELATIVE MOTION	36
RIVERBOAT PROBLEMS	38
MORE EXAMPLES RELATED TO KINEMATICS	40
FORCE	53
PROJECTILE MOTION	53
NEWTON'S LAWS OF MOTION	58
MORE PROBLEMS RELATED TO DYNAMICS	71
WORK	78
POWER	81
ENERGY	82
MORE PROBLEMS RELATED TO WORK POWER ENERGY	91
MOMENTUM	99
IMPULSE	100
CONSERVATION OF MOMENTUM	103
COLLISIONS	107
MORE EXAMPLES RELATED TO IMPULSE MOMENTUM	109
LINEAR SPEED (TANGENTIAL SPEED)	114
ANGULAR SPEED	116
ANGULAR ACCELERATION	117
CENTRIPETAL FORCE	118
CIRCULAR MOTION ON INCLINED PLANES	120
CENTRIFUGAL FORCE	121
TORQUE	122
MORE EXAMPLES RELATED TO ROTATIONAL MOTION	126
PROPERTIES of LIGHT	131

- REFLECTION of LIGHT 131
- PLANE MIRRORS AND IMAGE FORMATION IN PLANE MIRRORS 133
- CURVED MIRRORS 135
- REFRACTION 147
- CRITICAL ANGLE AND TOTAL REFLECTION 151
- TOTAL REFLECTION IN PRISMS 154
- APPARENT DEPTH REAL DEPTH 155
- MORE EXAMPLES RELATED TO OPTICS 157
- MATTER 167
- MASS 167
- INERTIA 167
- VOLUME 167
- DENSITY 168
- ELASTICITY 171
- MORE EXAMPLES RELATED TO PROPERTIES OF MATTER 172
- TEMPERATURE 177
- HEAT 179
- HEAT TRANSFER 180
- CHANGE OF PHASE/STATE 183
- PHASE TRANSITION OF WATER 188
- COMMON MISCONCEPTIONS 189
- THERMAL EXPANSION AND CONTRACTION 189
- MORE EXAMPLES RELATED TO HEAT TEMPERATURE AND THERMAL EXPANSION 194
- TYPES OF CHARGING 202
- GROUNDING 204
- Electroscope 205
- ELECTRICAL FORCES COULOMB'S LAW 209
- ELECTRIC FIELD 212
- FORCE ACTING ON A CHARGED PARTICLE INSIDE ELECTRIC FIELD 215
- ELECTRIC POTENTIAL AND ELECTRIC POTENTIAL ENERGY 217
- CAPACITANCE AND CAPACITORS 220
- MORE EXAMPLES RELATED TO ELECTROSTATICS 226
- ELECTRIC CURRENT AND FLOW OF CHARGE 234
- OHM'S LAW RESISTANCE AND RESISTORS 235
- COMMON ELECTRIC CIRCUITS AND COMBINATION of BATTERIES 240
- ALTERNATING CURRENT AND DIRECT CURRENT AND DIODES 244
- ELECTRIC POWER AND ENERGY 245
- FINDING POTENTIAL DIFFERENCE BETWEEN TWO POINTS IN CIRCUITS 247
- MORE EXAMPLES RELATED TO ELECTRIC CURRENT 249
- TYPES OF MAGNETS 256
- COULOMB's LAW FOR MAGNETISM 256
- MAGNETIC FIELD 258

MAGNETIC FLUX	260
MAGNETIC PERMEABILITY	261
MAGNETIC FIELD OF EARTH	262
MAGNETIC EFFECT OF THE CURRENT	263
FORCE ACTING ON MOVING PARTICLE AND CURRENT CARRYING WIRE	270
FORCE ACTING ON CHARGED PARTICLE	272
FORCES OF CURRENT CARRYING WIRES ON EACH OTHER	272
TRANSFORMERS	273
PROPERTIES OF WAVES	283
DIRECTION OF WAVE PROPAGATION	284
VELOCITY OF THE SPRING PULSE	284
VELOCITY OF PERIODIC WAVES	286
INTERFERENCE OF SPRING WAVES	287
REFLECTION OF SPRING WAVES	288
WATER WAVES	290
REFRACTION OF WAVES	292
MORE EXAMPLES RELATED TO WAVES	296

VECTOR AND SCALAR QUANTITIES

We can classify quantities under two groups: vector and scalar quantities. **Scalars** show only magnitude of that quantity. For example, time, temperature, mass, volume etc. As you can see, there is no mention on direction in scalar quantities. On the other hand, **vectors** are used for quantities, which have both magnitude and direction. For instance, velocity, acceleration, force, displacement are some of the examples of vector quantities. They all have both magnitude and direction. It is important us to know direction of velocity or acceleration, on the other hand in scalar quantities there is no need to know direction.

We will learn properties of vectors firstly and then pass to the vector quantities. You will be more familiar with most of the physics concepts after learning vectors. Look at the given shape, which shows parts of vector having both magnitude and direction.

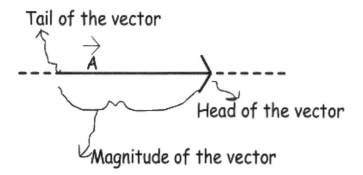

Head of the vector shows the direction and tail shows starting point. We can change position of the vector however, we should be careful not to change the direction and magnitude of it. In next subject we will learn how to add and subtract vectors. Moreover, we will learn how to find X and Y components of a given vector using a little bit trigonometry.

PROPERTIES OF VECTORS

Vectors can be summed with each other and result becomes new vector. We can multiply vectors with a scalar. If the scalar is positive, then only magnitude of the vector changes. However, if the scalar is negative then, both magnitude and direction of vector changes.

ADDITION OF VECTORS

We can add vectors by considering their vector properties. Vector, which is the summation of more than one vector is called **resultant vector** and symbolized with **R**. Look at the picture given below. It shows the classical addition of three vectors. We can add them just like they are scalars. However, you should be careful, they are not scalar quantities. They have both magnitude and direction. In this example their magnitudes and directions are the same thus; we just add them like scalars and write the resultant vector. Be careful, direction of resultant vector is same with the direction of A, B and C.

Let's look at a different example.

In this example, as you can see vector A has negative direction with respect to vectors B and C. So, while we add them we should consider their directions and we put a minus sign before the vector A. As a result, our resultant vector becomes smaller in magnitude than the first example but have same direction.

There are some key points you can use while finding resultant vector. Now, let's look at these special cases that can help us in solving problems.

A and B symbolize two different vectors and Θ symbolize angle between them and R is the resultant vector.

If A ≠ B in magnitude;

$\Theta = 0°$; $R = A + B$

$\Theta = 90°$; $R = \sqrt{(A^2 + B^2)}$

$\Theta = 180°$; $R = A - B$

If A = B in magnitude;

Θ = 0°; R = A + B = 2A

Θ = 60°; R = A√3

Θ = 90°; R = A√2

Θ = 120°; R = A

Θ = 180°; R = 0

Now lets solve some examples using these special cases.

Example: Find addition of vectors A, B, and C given below.

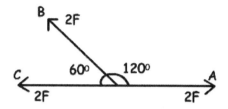

As you can see from the picture given above, vectors A and C have same magnitude but opposite direction. In other words, angle between A and C is 180^0, thus their resultant vector becomes zero. So, we have only vector B which becomes resultant vector of A, B and C.

MULTIPLYING A VECTOR WITH A SCALAR

When we multiply a vector with a scalar quantity, if the scalar is positive than we just multiply scalar with magnitude of the vector. But, if the scalar is negative then we must change the direction with magnitude of the vector. Example given below shows the details of multiplication of vectors with scalar.

Example: Find 2A, -2A and 1/2A from the given vector A.

⟶   A=4

⟶   2A=8

⟵   -2A=-8

⟶   1/2A=2

As you can see, when we multiply a vector with a positive scalar its direction does not change. On the contrary when we multiply a vector with a negative scalar its direction changes.

COMPONENTS OF VECTORS

Vectors are not given all the time in four directions. To make calculations simpler sometimes we need to show vectors in X, -X and Y, -Y components. For example, look at the vector given below. It is in northeast direction. In the figure, we see X and Y components of this vector. In other words, addition of Ax and Ay gives us vector A.

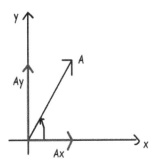

We benefit from trigonometry at this point. I will give two simple equations and you can use them to find components of any given vector.

$$\sin 60° = \frac{Ay}{A} \quad \text{and,} \quad \cos 60° = \frac{Ax}{A}$$

Thus,

$$Ax = A \cdot \cos 60°$$
$$Ay = A \cdot \sin 60°$$

All vectors can be divided into their components. Now we solve an example and see how we use this technique.

Example: Find the resultant vector of A and B given in the graph below. ($\sin 30° = 1/2$, $\sin 60° = \sqrt{3}/2$, $\sin 53° = 4/5$, $\cos 53° = 3/5$)

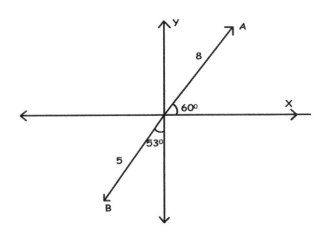

We use trigonometric equations to find components of the vectors. Then, make addition and subtraction between vectors sharing same direction to find resultant vector.

Components of A:
$A_x = A \cdot \cos 60°$
$A_x = 8 \cdot 1/2 = 4$
$A_y = A \cdot \sin 60°$
$A_y = 8 \cdot \frac{\sqrt{3}}{2} = 4\sqrt{3}$

Components of B:
$B_x = B \cdot \sin 53°$
$B_x = 5 \cdot 4/5 = 4$
$B_y = B \cdot \cos 53°$
$B_y = 5 \cdot 3/5 = 3$

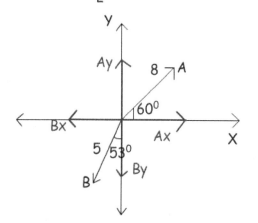

We sum the vectors having same direction:

$A_x + B_x = 4 - 4 = 0$

$A_y + B_y = 4\sqrt{3} - 3$

We put "-" in front of B_x, and B_y because we take right side and upward direction as positive

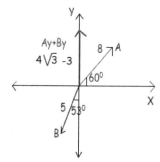

MORE PROBLEMS RELATED TO VECTORS

Example: Use following picture and find summation of given vectors A+B+C.

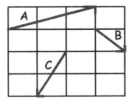

First, we find A+B and then add it to vector C.

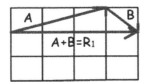

We find R_1, now we add C to R_1 to find resultant vector.

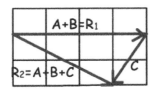

R_2 = A+B+C

Example: There are five vectors A, B, C, D and E in the given picture. Find resultant vector of them.

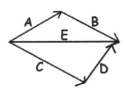

Since A+B = E and C+D = E

R = A+B+C+D+E

R = E+E+E = 3E

Example: If A and A+2B vectors are like given below, find vector B.

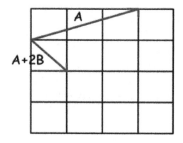

We use vector addition properties.

A + 2B – A = 2B

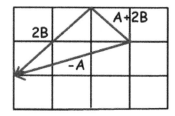

To get vector B, we multiply 2B with 1/2.

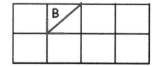

Example: Find resultant vector in the picture given below.

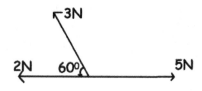

$F_1 + F_2 = 5 - 2 = 3N$

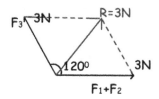

$F_1 + F_2 + F_3 = R = 3N$

Example: Which one of the following statements is true about the vectors given below?

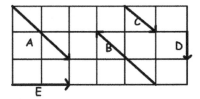

I A = B in magnitude

II A = 2C

III E = 2D

IV A = B

As you can see in the figure given above, A and B are equal in magnitude, so I is true. If you multiply C with 2, you get A, this means that II is also true. E = 2D in magnitude but not in direction. Thus, III is false. A = B in magnitude but not in direction so IV is false.

Example: Resultant vector of K, L and M is zero.

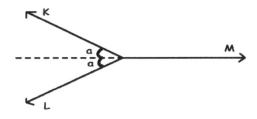

Which one of the following statements given below is definitely false?

I. Ky and Ly components are equal vectors

II. K + L = M

III. a = 60^0

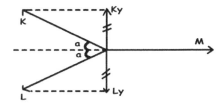

I. Ky = -Ly, they are equal in magnitude but opposite in directions. Thus they are not equal vectors, I is false.

II. Magnitude of K + L = M, but directions are opposite so II is false also.

III. a = 60^0 is possible. III is not exactly false.

Example: If $α_3 < α_2 < α_1$ and $R_1 = R_2 = R_3$, find the relation between F_1, F_2 and F_3.

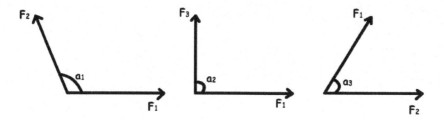

Decreasing in the angle between forces increases the resultant force.

If $α_1 = α_2$, then $R_1 > R_2$ and $F_2 > F_3$

If $α_2 = α_3$, then $R_2 > R_3$ and $F_1 > F_2$

$F_1 > F_2 > F_3$

KINEMATICS

In the last section we have learned scalar and vector concepts. Beyond the definitions of these concepts in kinematics we will try to explain speed, velocity, acceleration, distance and displacement, free fall and relative motion terms. As mentioned in last section distance and displacement are different terms. Distance is a scalar quantity and displacement is a vector quantity. In the same way we can categorize speed and velocity. Speed is a scalar quantity with just concerning the magnitude and velocity is a vector quantity that must consider both magnitude and direction. Let me explain speed and velocity in detail one by one.

DISTANCE AND DISPLACEMENT

Distance is a scalar quantity representing the interval between two points. It is just the magnitude of the interval. However, Displacement is a vector quantity and can be defined by using distance concept. It can be defined as distance between the initial point and final point of an object. It must be the shortest interval connecting the initial and final points, that is a straight line. Let's look at the below examples for deep understanding.

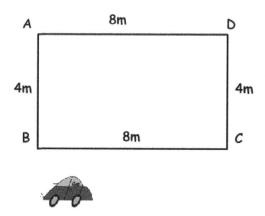

Look at the picture above, car travels from D to A, A to B, B to C and C to D. Displacement from D to D (which are our initial and final points) is zero. However, distance traveled is not zero. It is equal to the perimeter of the rectangle.

Example: Look at the picture given below. An object moves from point A through B, C, D, E and stops at point F.

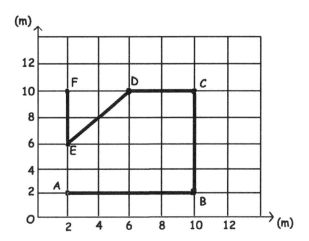

a) Find final displacement.

b) Find distance taken from point A to D.

a) We find final displacement by drawing straight line from point A to final point F. As you can see from the graph, object changes its position 8m.

Displacement = Final position - Initial position

Displacement = 10m - 2m = 8m

b) We find distance taken by object;

A to B = 10 - 2 = 8m

B to C = 10 - 2 = 8m

C to D = 10 - 6 = 4m

Total distance taken from point A to D is = 8m + 8m + 4m = 20m

SPEED

Speed can be defined as "how fast something moves" or it can be explained more scientifically as "the distance covered in a unit of time". In daily life we use the first definition and say the faster object has higher speed. Speed does not show us the direction of the motion it just gives the magnitude of what distance taken in a given time. In other words it is a scalar quantity. We use a symbol v to show speed. Let me formulate what we talk above;

Speed = distance/time

slower

faster

From the above formula we can say that speed is directly proportional to the distance and inversely proportional to the time. I think it's time to talk a little bit the units of speed. Motor vehicles commonly use kilometer per hour (km/h) or mile per hour (mph) as a unit of speed however in short distances we can use meter per second (m/s) as a unit of speed. In my examples and explanations I will use m/s as a unit.

Example: Calculate the speed of the car that travels 450m in 9 seconds.

Speed = Distance/Time

Speed = 450m/9s

Speed = 50m/s

VELOCITY

Velocity can be defined as "speed having direction". As you can understand from the definition velocity is a vector quantity having both magnitude and direction. In daily life we use speed and velocity interchangeably but in physics they have different meanings. We can define velocity as the "rate of change of displacement" whereas "the speed is rate of change of distance". While we calculate speed we look at the total distance covered in a given time, however, in calculating velocity we must consider the direction and in short we can just look at the change in position not the whole distance traveled. If a man walks 5m to east and then 5m to west speed of that man calculated by dividing total distance traveled which is 10m to the time elapsed, however, velocity calculated by dividing the displacement to the elapsed time, which is 0m divided elapsed time gives us zero. In other words, if the displacement is zero velocity becomes zero.

❗ Be careful!! There must be a change in the position of the object to have a velocity.

We use Δ symbol to show the change in something. For example, we can symbolize the change in position as ΔX.

Example: Calculate the speed and velocity of the man moving 45m to the north, and 36m to the south in 27 seconds.

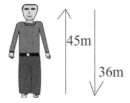

First we should calculate distance traveled and displacement of the man to calculate speed and velocity.

Total distance covered = 45m + 6m = 81m

Speed = total distance/time of travel = 81m / 27s = 3m/s

Velocity = displacement/time = (45-36) m / 27s = 9m /27s = 0,33m/s

We show with this example that, speed and velocity are not same thing.

AVERAGE SPEED AND INSTANTANEOUS SPEED

A moving object does not have the same speed during its travel. Sometimes it speeds up and sometimes slows down. At a given instant time what we read from the speedometer is **instantaneous speed**. On the other hand average speed should be calculated differently. For example, a car moving with a constant speed travels to another city, it must stop at red lights in the traffic, or it should slow down when unwanted situations occur in the road.

At the end of the trip, if we want to learn **average speed** of the car we divide total distance to total time the trip takes.

$$AverageSpeed = \frac{TotalDistanceTravelled}{TimeInterval} \quad AverageSpeed = \frac{TotalDistanceTraveled}{TimeInterval}$$

Assume that car travels 500 km in a 5 hour. When we calculate average speed we see that it is 100km/h. Of course the car does not travel with a 100 km/h constant speed. It has many instantaneous speeds and 100 km/h is the average of those instantaneous speeds.

AVERAGE VELOCITY AND INSTANTANEOUS VELOCITY

We can follow the same steps used in the definition of average and instantaneous speed while defining average and instantaneous velocity. Instantaneous velocity is the velocity at a given instant of time, however, as in the case of velocity, average velocity is calculated with displacement over time interval.

$$Average\ Velocity = \frac{Displacement}{Time\ Interval}$$

Example: A man is traveling with his car 150m to the east and 70m to the west; calculate the average speed and velocity of the car if the duration of travel is 10 s.

Average Velocity=Displacement/Time Interval   Displacement=150m-70m=80m

Average Velocity= 80m/10s=8m/s east

Average Speed=Total Distance Traveled/Time Interval

Average Speed= (150m+70m)/10s

Average Speed=22m/s

This is a good example, which shows the difference of velocity and speed clearly. We must give the direction with velocity since velocity is a vector quantity however, speed is a scalar quantity and we do not consider direction.

ACCELERATION

Definition of acceleration is a little bit different from speed and velocity. We can easily define **acceleration** as "change in velocity". As you understood from the definition there must be change in the velocity of the object. This change can be in the magnitude (speed) of the velocity or the direction of the velocity. In daily life we use **accelerating** term for the speeding up objects and **decelerating** for the slowing down objects. I want you to focus on here! In physics we use acceleration concept a little bit different from its daily life usage. If there is a change in the velocity whether it is slowing down or speeding up, or changing its direction we say that object is accelerating.

The mathematical representation of acceleration is given below. "a" is acceleration, "v" is velocity and "t" is time.

$$Acceleration = \frac{Change\ in\ Velocity}{Time\ Interval}$$

Or;

$$a = \frac{V_f - V_i}{t}$$

In figure given below you see a car moving in a curved path. While it is traveling its direction is changing with the path. However, we cannot observe if there is a change in its speed or not. Can we say that the car has acceleration?

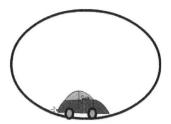

The answer of the question is yes. In the definition of acceleration we have said that for having acceleration there must be change in the magnitude or direction of the velocity. In this example we see that the direction of the car also velocity is changing with time. Thus, of course we can say that this car is accelerating.

Now better understanding lets solve some problems.

Example: A car starts to move and reaches the velocity 80m/s in 10 seconds. Calculate the acceleration of this car?

$$a = \frac{V_f - V_i}{t}$$

$$a = \frac{80 m/s - 0 m/s}{10s} = \frac{80 m/s}{10s} = 8 m/s/s = 8 m/s^2$$

Example: A boy starts to run with 3m/s² acceleration. Calculate the boy's final velocity after 15 seconds?

$a = 3 m/s^2$
$t = 15s$

$V_f = ?$

Acceleration = (Change in Velocity) / (Time Interval)

Change in Velocity = Acceleration x Time

$V_f - V_i = 3 \text{ m/s}^2 \times 15s$

$V_f - 0 = 45 \text{ m/s}$

$V_f = 45 \text{ m/s}$

Common Misconceptions

• There are some important points related to speed, velocity and acceleration that you should focus on carefully. Most of the students have some misconceptions about this subject. Some of them are given below.

• They think that if the velocity is constant than acceleration is also constant. However, as we discussed before, if there is no change in velocity in other words if velocity is constant both in magnitude and direction than the acceleration is zero.

• Another misconception is that if speed of the object is constant than acceleration is zero. As I mentioned before, speed only shows us the magnitude of the velocity. We cannot ignore its direction. Thus, acceleration can be zero or not. There is only one option making acceleration zero when speed is constant. If the object does straight line motion with a constant speed then we say acceleration is zero. On the contrary if the object does circular motion with a constant speed we cannot talk about zero acceleration because; in this situation direction of the velocity is changing.

• They think that if the object moves with a high velocity then its acceleration is also high and if the object moves with a low velocity then the acceleration is also low. But, high velocity does not mean high acceleration. Acceleration can be high, low or zero in high velocities. It is only related to the change in velocity. If the change in velocity high then the acceleration will be high if the change in velocity is low then the acceleration is low.

• Most of the students believe that if the acceleration is positive than the object speeds up, and if the acceleration is negative then the object is slowing down. We cannot conclude the motion of the object just looking at its acceleration. We should consider the sign of the velocity while talking about whether it is speeding up or slowing down. If the acceleration is positive and velocity is also positive then we can say that object is speeding up in the positive direction. Moreover, if the acceleration is negative and velocity is also negative we say object is speeding up in negative direction. On the contrary if the velocity and acceleration has opposite directions such as acceleration is positive and velocity is negative or vice versa we say object is slowing down. In summary, to decide if the object is speeding up or slowing down we must look at both directions of the velocity and acceleration.

• Sometimes negative acceleration makes confusion in students' mind. This negative symbol has physical meaning. We use it to show the direction of the vector quantities.

DESCRIBING MOTION WITH GRAPHS

Position vs. Time Graphs

Graphs are commonly used in physics. They give us much information about the concepts and we can infer many things. Let's talk about this position vs. time graph. As you see on the graph, X-axis shows us time and Y axis shows position. We observe that position is linearly increasing in positive direction with the time. We understand from this linear increasing, our **velocity is constant**. If it is not constant, we will see a curved line in our graph.

1. Position vs. time graph in which position increases linearly with time.

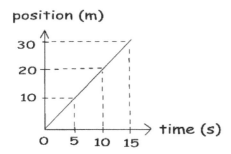

Now, we use this graph and make some calculations.

$$v = \frac{\text{Change in position}}{\text{time interval}} \qquad v = \frac{V_f - V_i}{\text{time interval}}$$

$$v = \frac{30m - 0m}{15s - 0s} \qquad v = 2 \, m/s$$

From the given graph we calculate velocity; there is another way to make this calculation. We just look at slope of the graph and find velocity. What we mean by slope is;

$$\text{Slope} = \frac{X_f - X_i}{t_f - t_i}$$

Which is the equation we use in calculation of the velocity.

2. In this graph, position increases parabolically with time in positive direction.

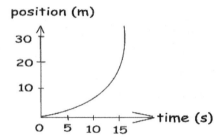

In this graph our velocity is changing. As a result of this change graph has curved line not linear. So, position does not increase linearly. We can also find the velocity of the object from this graph. We should first find the slope of the curve and calculate the velocity.

Example: Using the given graph find the velocity of the object in intervals (1s – 3s) and (3s – 5s).

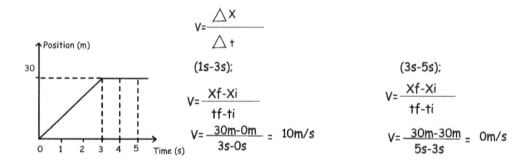

$$V = \frac{\Delta x}{\Delta t}$$

(1s-3s);
$$V = \frac{X_f - X_i}{t_f - t_i}$$
$$V = \frac{30m - 0m}{3s - 0s} = 10 m/s$$

(3s-5s);
$$V = \frac{X_f - X_i}{t_f - t_i}$$
$$V = \frac{30m - 30m}{5s - 3s} = 0 m/s$$

While solving graph problems you should be careful while reading them. In this example, at interval (3s-5s) position does not change. You can easily see it from the graph, however, lets prove it with some calculations. If there is no change in position then there is no velocity or vice versa. You can say more things about the motion of the object by just looking at the graph. The important thing is that you must know the relations, meaning of the slopes or area under the graphs. We will solve more examples using graph for deep understanding and analyze the motion from the graphs.

3. This position time graph is an example of increasing position in negative direction.

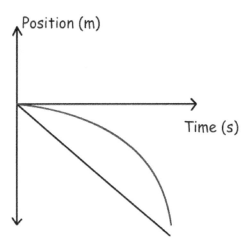

Curved line shows nonlinear increasing and linear line shows linear increasing in position in negative direction. We say that linear increasing in position is the result of constant velocity that means zero acceleration. Moreover, nonlinear increase in the position is the result of changing velocity and it shows there is a nonzero acceleration.

Until now we saw graphics including speeding up motion in positive and negative directions. Now we talk about a little bit slowing down object's graphs. For example, in the given graph below, our position is decreasing in different directions. Curved line in lower side of graph shows the position of the object, which is slowing down, in negative direction and curved line in upper side of the graph shows the position of the object slowing down in positive direction. However, the linear lines in the upper and lower side of the graph show the linear decrease in position in positive and negative directions.

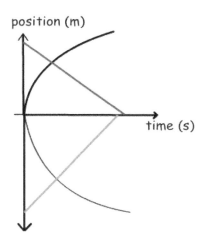

We have seen various type of position vs. time graphs. I think they will help you in solving problems. It is really easy, you should just keep in mind that "slope of position-time graph shows the velocity".

Velocity vs. Time Graphs

In velocity vs. time graphs, X-axis is time as in the case of position vs. time graphs and Y-axis is velocity. We can benefit from this graph by two ways. One of them is area under the graph, which gives the displacement and the slope, which gives the acceleration.

We have talked about different kinds of motion such as, constant motion having constant velocity, accelerated motion like speeding up or slowing down. For instance, in this graph as it seen our velocity is constant, time passes but velocity does not change. What you see when you look at the graph given below?

1.

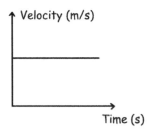

We learn relation between time and velocity, how velocity is changing with time. In this graph we can say that acceleration is zero since velocity is constant. Moreover, using velocity vs. time graphs we can calculate displacement of the object. How can we do this? Let's think together. First, look at the definition of displacement;

Displacement = Velocity x Time

Using formula given above, we can say that **area** under velocity vs. time graph also gives us displacement of the object. Look at the example given below to understand what we mean by "area under the graph".

Example: Use velocity vs. time graph given below and calculate displacement of the object for the interval (0s – 4s).

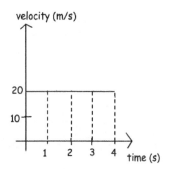

To solve this problem, I use the technique given above and classic formula. The area under the graph will give us the displacement. Then we compare the results of two techniques.

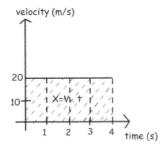

Area of the rectangle= A X B

where A and B are the sides of the rectangle

Area of the rectangle= 20m/s X 4s =80m

Displacement=Velocity X time

Displacement=20m/s X 4s=80m

As you can see, the results are same, thus, we can say that in velocity vs. time graphs we can find the displacement by looking at area under the graph.

2. In this graph there is a linear increase in velocity with respect to time so; the acceleration of the motion is constant. Moreover, we can also calculate the displacement by looking at under the area of the graph.

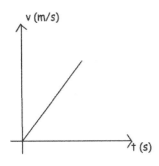

Let's solve another example for deep understanding.

Example: Calculate displacement of the car from the given graph.

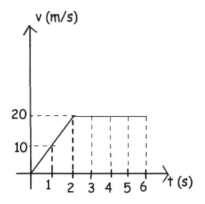

We can calculate displacement by using area under velocity vs. time graph; to do this we can first calculate the area of the triangle and then rectangle. Finally the sum of these two areas gives us the total displacement of the car.

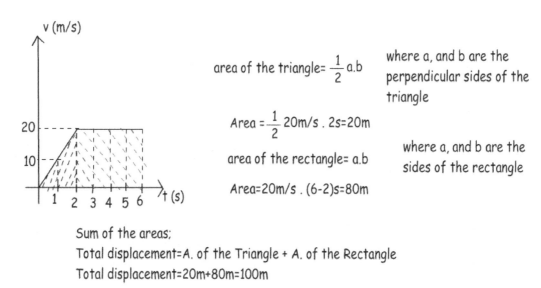

area of the triangle = $\frac{1}{2}$ a.b   where a, and b are the perpendicular sides of the triangle

Area = $\frac{1}{2}$ 20m/s . 2s = 20m

area of the rectangle = a.b   where a, and b are the sides of the rectangle

Area = 20m/s . (6-2)s = 80m

Sum of the areas;
Total displacement = A. of the Triangle + A. of the Rectangle
Total displacement = 20m + 80m = 100m

3. The graph given below shows different accelerated motions. The lines are curved because acceleration is not constant. They represent the decreasing and increasing velocity in positive and negative directions. However, we do not deal with such problems now.

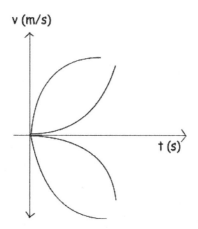

⚠ We have discussed so far velocity vs. time graphs. When you look at these graphs you can find the displacement and acceleration of the object by looking at the **slope for acceleration and area for displacement.**

FREE FALL

Free fall is a kind of motion that everybody can observe in daily life. We drop something accidentally or purposely and see its motion. At the beginning its speed is zero and until the end it gains speed and before the crash it reaches its maximum speed. Which factors affect speed of the object, while it is in free fall? How can we calculate the distance and time it takes during free fall? We deal with these subjects in this section. First, let me begin with source of increasing in amount of speed during the fall.

As you can guess, things fall because of gravity. Thus, our objects gain speed approximately 10m/s in a second while falling because of the gravitation. We call this acceleration in physics **gravitational acceleration** and show with "g". The value of g is 9,8m/s² however; in our examples we assume it 10m/s² for simple calculations. Now it's time to formulate what we said above. We talked about the increase in speed, which is equal to the amount of g in a second. Thus our velocity can be found by the formula;

V=g.t   where; g is gravitational acceleration and t is the time.

Look at the given example below and try to understand what I tried to teach above.

Example: The boy drops the ball from a roof of the house, which takes 3 seconds to hit the ground. Calculate the velocity before the ball crashes to the ground. (g=10m/s²)

Velocity is;

V=g.t

V=10m/s².3s=30m/s

We have learned how to find the velocity of the object at a given time. Now we will learn how to find the distance taken during the motion. I give some equations to calculate distance and other quantities. Galileo found an equation for distance from his experiments.

This equation is;

$$\text{Distance Traveled} = \frac{1}{2} g.t.t = \frac{1}{2} g.t^2$$

Using this equation we can find the height of the house in given example 7. Let's found how height the ball has been dropped? We use 10 m/s² for g.

$$\text{Distance Traveled} = \frac{1}{2} g \cdot t \cdot t = \frac{1}{2} \cdot 10 \cdot 3^2$$

Distance Traveled = 45m

I think, with the help of given calculations, formula becomes clearer in your minds. We will solve more problems related to this topic. Now, think that if I throw the ball straight upward with an initial velocity. When it stops and falls back to the ground? We answer these questions now.

Picture shows magnitudes of velocity at the bottom and at the top. As you can see ball is thrown upward with an initial velocity "v", at the top its velocity becomes zero and it changes its direction and starts to fall down that is called "free fall". Finally at the bottom before the crash it reaches its maximum speed which shown as "V". We have talked about amount of increase in velocity in free fall. It increases 9,8m/s in each second due to the gravitational acceleration. In this case, there is also g but the ball's direction is upward; so the sign of g is negative. Thus, our velocity decreases in 9,8m/s in each second until the velocity becomes zero.

At the top, since velocity is zero, ball changes its direction and starts to free-fall. Before solving problems I want to give graphs of free fall motion.

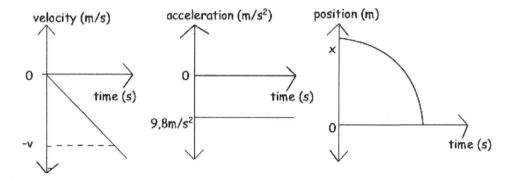

As you see in first graph given above, our velocity increases linearly with acceleration "g", second graph tells us; acceleration is constant at 9,8m/s², and finally third graph is the representation of change in position. At the beginning we have a positive displacement and as the time passes it decreases and finally becomes zero. Now we can solve problems using these graphs and explanations.

Example: John throws the ball straight upward and after 1 second it reaches its maximum height then it does free fall motion which takes 2 seconds. Calculate maximum height it reaches and velocity of the ball before it crashes the ground. (g=10m/s²)

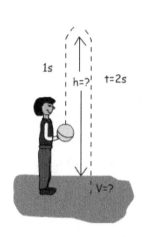

$V = g \cdot t$
$V = g \cdot t = 10 m/s^2 \cdot 1s = 10 m/s$ ball is thrown with 10 m/s velocity

at the top our velocity is zero,
ball does free fall

$V = -g \cdot t$
$V = -g \cdot t = -10 m/s^2 \cdot 2s = -20 m/s$ we put "-" sign in front of the g because we take upward direction "+"

$\text{Distance} = \dfrac{1}{2} g \cdot t^2$

$h_{max} = \dfrac{1}{2} \cdot 10 m/s^2 \cdot (2s)^2$

$h_{max} = 20 m$

Example: An object does free fall motion. It hits the ground after 4 seconds. Calculate the velocity of the object after 3 seconds and before it hits the ground. What can be the height it is thrown?

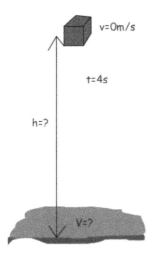

Velocity after 3 second is;
V = -g.t
V = -10m/s².3s = -30 m/s
Velocity after 4 second is;
V = -10m/s².4s = -40 m/s
Height is;
h = 1/2 . g . t² = 1/2 . 10m/s².(4s)²
h = 80 m

Two examples given above try to show how to use free fall equations. We can find the velocity, distance and time from the given data. Now, I will give three more equations and finish 1D Kinematics subject. The equations are;

$$V = V_0 + at$$

$$X = V_0 t + \frac{1}{2} . a . t^2$$

$$V^2_f = V^2_i + 2.a.X$$

First equation is used for finding velocity of the object having initial velocity and acceleration. Second one is used for calculating distance of the object having initial velocity and acceleration. Third and last equation is timeless velocity equation. If distance, initial velocity and acceleration of the object are known then you can find the final velocity of the object. Now let's solve some problems using these equations to comprehend the subject in detail.

Example: Calculate velocity of the car, which has initial velocity 24m/s and acceleration 3m/s² after 15 second.

We use the first equation to solve this question.

$V = V_0 + a \cdot t$
$V = 24 m/s + 3 m/s^2 \cdot 15 s$
$V = 69 m/s$

Example: The car that is initially at rest has an acceleration 7m/s² and travels 20 seconds. Find the distance it covers during this period.

$X = V_0 t + \frac{1}{2} \cdot a \cdot t^2$
$X = \frac{1}{2} \cdot 7 m/s^2 \cdot (20s)^2$
$X = 1400 m$

RELATIVE MOTION

When we talk about velocity of something we first determine a reference point and then according to this reference point we say velocity of the object. For example, you are in a bus and it goes with velocity of 50 m/s to the east, then a truck passes you with a velocity of 60m/s to the east. When truck is next to the bus you feel that as you go backward to the west. In other example, when two cars have same velocity and if you look another car you feel you do not move. On the contrary, if you look at stationary objects at the ground when you travel, then you feel yourself travel with the velocity of your car.

All these examples are result of relative motion. Velocity of the moving objects with respect to a reference point is called "**relative velocity**" and this motion is called "**relative motion**". Reference point is too important in physics. We do all calculations according to the reference point. For instance, we observe plane flying in the air, velocity of that plane with respect to us is the sum of the velocities of plane and wind. However, the same plane has different velocity with respect to other flying plane. To sum up, we determine directions and quantities of velocity of the objects with respect to a chosen reference point. Now we look at some examples and vector notations of relative velocity of the objects.

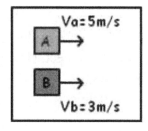 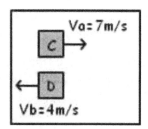

In the first picture according to the observer in A, B is traveling with velocity of 2m/s to the west. However, according to the observer in B, A is traveling with velocity of 2m/s to the east.

In the second picture observer in C feels that as if they are traveling with a velocity of 11m/s to the east with respect to D. On the contrary, observer in D feels that as if they are traveling with a velocity of 11m/s to the west with respect to C.

Now, I give you an equation that you can easily find relative velocities of the objects with respect to a reference point.

V (relative) = V (object) - V (observer)

Example: Find velocity of B with respect to A.

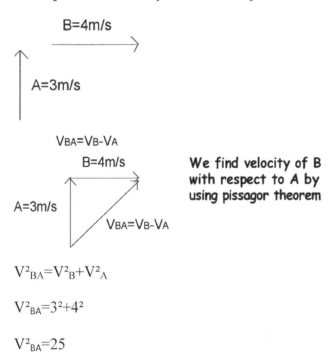

We find velocity of B with respect to A by using pissagor theorem

$V^2_{BA} = V^2_B + V^2_A$

$V^2_{BA} = 3^2 + 4^2$

$V^2_{BA} = 25$

$V_{BA} = 5$m/s to the southeast

Example: Find the velocity of L with respect to K and M.

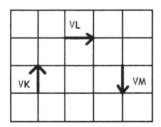

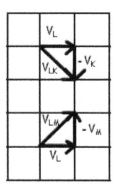

We use diagram given above to find relative velocities of L.

$V_{LK} = V_L - V_K$ and $V_{LM} = V_L - V_M$

RIVERBOAT PROBLEMS

Riverboat problems are basic examples of relative motion. You can solve any kind of river problem by using principles of relative motion. In this section, we solve different types of river problems to make relative motion clearer in your mind.

Example: Velocity of boat with respect to river is 5 m/s. It aims to reach point A, however, it reaches opposite shore at point B because of the river velocity. If the velocity of current is 2m/s to the east calculate the time of trip and distance between A and B.

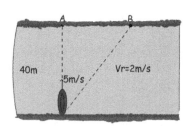

Picture given below shows the path of boat (dashed line), which is resultant vector. This example can be examined under two part, vertical and horizontal motion as in the case of projectile motion.

Vertical Motion
X=V.t
40m=5m/s.t
t=8s
trip takes 8s.

where, X is the width of the river, V is the speed of boat with respect to river and t is the time trip takes.

Horizontal Motion

X=V.t   where, X is the distance between points A and B, V is the speed of river and t is the time.
X=2m/s.8s
X=16m

Distance between A and B is 16m

Example: Velocity of the boat with respect to river is 10 m/s. It passes river and reaches opposite shore at point C. If velocity of the river is 3m/s, find the time of trip and distance between B and C.

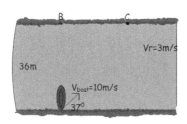

First we find the components of V_{boat}.

$V_x = V_{boat} \cdot \cos 37°$
$V = 10m/s \cdot 0{,}8 = 8m/s$
$V_y = V_{boat} \cdot \sin 37°$
$V_y = 10m/s \cdot 0{,}6 = 6m/s$

Vertical Motion

$X = V_{boaty} \cdot t$ We use the y component of boat's velocity
36m=6m/s.t
t=6s

Horizontal Motion
We must find the resultant velocity in direction

$\longrightarrow$ + $\longrightarrow$ = $\longrightarrow$
$V_{boatyx}=8m/s$ $V_{river}=3m/s$ $V_{resultant}=11m/s$

$X = V_{resultant} \cdot t$
X=11m/s.6s
X=66m   Distance between B and C is 66m

MORE EXAMPLES RELATED TO KINEMATICS

Example: Velocity vs. time graph of an object traveling along a straight line given below.

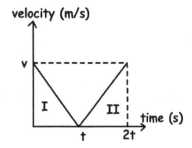

Draw the acceleration vs. time and position vs. time graph of the object by using graph given above.

Slope of the velocity vs. time graph gives us acceleration. In first interval, slope of the line is constant and negative, thus, acceleration of the object is also constant and negative. In other words, object does slowing down motion in positive direction with negative acceleration.

Slope=(0-v)/t=-a

In the second interval, slope is constant and positive, so acceleration is also constant and positive. Object does speeding up motion in positive direction.

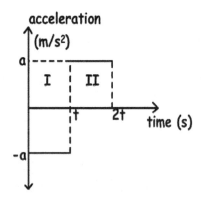

Slope=(v-0)/t=+a

In the first and second interval velocity of the object changes constantly thus; position time graph becomes like in the picture given below.

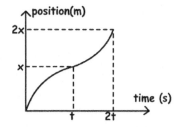

Example: An object is stationary at t=0. Picture given below shows the acceleration vs. time graph of this object. Find the intervals in which object speeds up in positive direction.

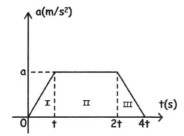

Area under the acceleration vs. time graph gives us the change in velocity.

In first interval, acceleration is increasing with the time, thus object speeds up with increasing acceleration.

In second interval, acceleration of the object is constant, so object speeds up constantly.

In the third interval, acceleration of the object is decreasing with the time, so object speeds up with decreasing acceleration.

In all intervals, velocity of the object increases.

Example: An object is stationary at t=0. Picture given below shows the acceleration vs. time graph of this object. Find the intervals in which object speeds up in positive direction.

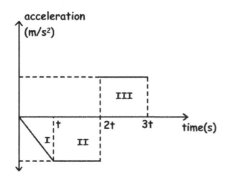

In the first interval, object speeds up with increasing acceleration in negative direction. In the second interval, object speeds up with constant acceleration in negative direction and in the third interval, object slows down with constant acceleration. (Acceleration (+), velocity (−))

Example: Two car **A** and **B** are at the same point at t=0. **B** travels with constant velocity and **A** travels with constant acceleration. Their velocity vs. time graph is given below. After how many meters they become together.

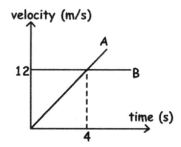

They have to take same distances to become together. Thus, using the graph given below we find distance they take. Area under the velocity vs. time graph gives us distance taken by the cars. We say that they become together after t s. Thus, we draw following graph in which triangle AOt shows distance taken by car A and rectangle B12Ot shows distance taken by car Y.

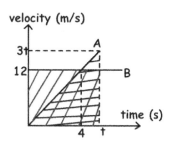

$a_x = 12/4 = 3 m/s^2$

Ax=Bx

3t.t/2=12.t

3t=24

t=8s

Bx=12.8=96m

Example: Position vs. time graph of a car is given below. In which intervals direction of velocity and direction of acceleration are same.

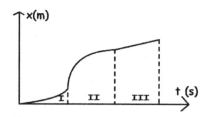

We draw velocity vs. time graph using position time graph.

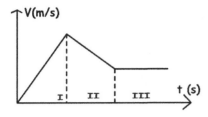

Slope of the velocity vs. time graph gives us acceleration.

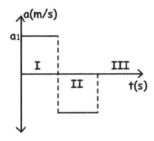

In the first interval, directions of velocity and acceleration are same.

Example: An object is dropped from 320 m high. Find the time of motion and velocity when it hits the ground.($g=10m/s^2$)

h=1/2.g.t^2 , v=g.t

h=320m

g=10m/s^2

320=1/2.10.t^2

t=8s.

v=g.t=10.8=80m/s

Example: An object does free fall and it takes 60m distances during last 2 seconds of its motion. Find the height it is dropped.($g=10m/s^2$)

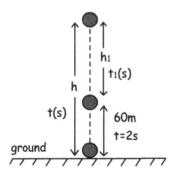

t is the time of motion

$h = 1/2 \cdot g \cdot t^2$

$h_1 = 1/2 \cdot g \cdot t_1^2$

put $t_1 = t-2$ and $h-h_1 = 60$ in the equation,

$1/2 \cdot g \cdot t^2 - 1/2 \cdot g \cdot t_1^2 = 60$

$5t^2 - 5(t^2 - 4t + 4) = 60$

$t = 4s$

$h = 1/2 \cdot g \cdot t^2 = 1/2 \cdot 10 \cdot 4^2 = 80m$

Example: An object is dropped from height of 144m and it does free fall motion. Distance it travels and time of motion are given in the picture below. Find the distance between points B-C.

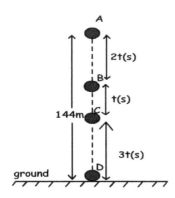

We can draw velocity time graph of object and area under this graph gives us position of the object.

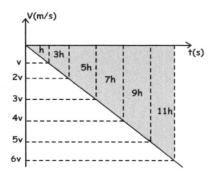

As you can see from the velocity time graph, object travels 5h distance during 2t-3t, which is the distance between points B and C.

All distance traveled is 36h

144m=36h

h=4m

Distance between B-C=5h=5.4m=20m

Example: A hot-air balloon having initial velocity v_0 rises. Stone dropped from this balloon, when it is 135 m height, hits the ground after 9 s. Find the velocity of the balloon.

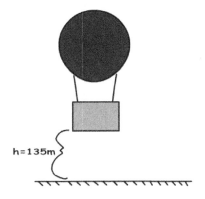

$-h = v_0 \cdot t_{fligth} - 1/2 \cdot g \cdot t_{fligth}^2$

$-135 = v0.9 - 1/2.10.(9)^2$

$-135 = 9v_0 - 405$

$9v_0 = 270$

$v_0 = 30 \text{m/s}$

Example: Look at the given picture below. Object K does free fall motion and object B thrown upward at the same time. They collide after 2s. Find the initial velocity of object B. ($g=10\text{m/s}^2$)

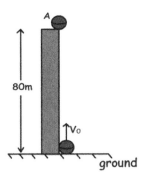

Object A does free fall motion

$h_A = 1/2.10.2^2 = 20\text{m}$

$h_L = v_0 \cdot t - 1/2 \cdot g \cdot t^2$

$h_L = v_0.2 - 1/2.10.2^2$

$h_L = 2v_0 - 20$

$h_K + h_L = 80m$

$20m + h_L = 80m$

$2v_0 - 20 = 60m$

$v_0 = 40m/s$

Example: As you can see in the given picture, ball is thrown horizontally with an initial velocity. Find the time of motion. ($g=10m/s^2$)

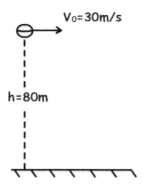

Ball does projectile motion in other words it does free fall in vertical and linear motion in horizontal. Time of motion for horizontal and vertical is same. Thus in vertical;

$h = 1/2g.t^2$

$80 = 1/2.10.t^2$

$t = 4s$

Example: An object hits the ground as given in the picture below. Find the initial velocity of the object.

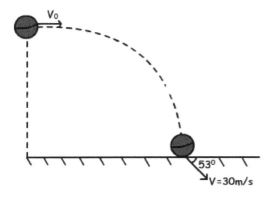

Velocity of horizontal motion is constant. So;

$V_0 = V_x = V\cos 53^0$

$V_x = V_0 = 30 m/s \cdot 0,6$

$V_0 = V_x = 18 m/s$

Example: An object is thrown with an angle 37^0 with horizontal. If the initial velocity of the object is 50m/s, find the time of motion, maximum height it can reach, and distance in horizontal.

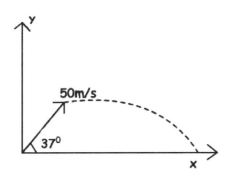

$V_{0x} = V_0 \cos 53^0 = 50 \cdot 0,8 = 40 m/s$

$V_{0y} = V_{0y} \cdot \sin 53^0 = 50 \cdot 0,6 = 30 m/s$

a) $V - V_{0y} = 0 - g \cdot t$ at the maximum height

$t = 30/10 = 3s$

$2 \cdot t =$ time of motion $= 2 \cdot 3 = 6s$

b) $V_{0y}^2 = h_{max} \cdot 2 \cdot g$

$h_{max} = 30^2/2 \cdot 10 = 45m$

c) $X = V_{0x} \cdot t_{total} = 40 \cdot 6 = 240m$

Example: A balloon having 20 m/s constant velocity is rising from ground to up. When the balloon reaches 160 m height, an object is thrown horizontally with a velocity of 40m/s with respect to balloon. Find the horizontal distance traveled by the object.

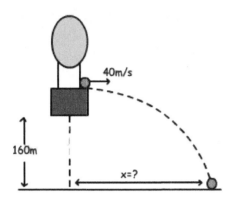

Object has velocity 40m/s in horizontal, 20m/s in vertical and its height is 160m. We can find time of motion with following formula;

$h = V_{0y}.t - 1/2.g.t^2$

$-160 = 20.t - 1/2.10.t^2$

$t^2 = 4t - 32$

$(t-8).(t+8) = 0$

$t = 8s$

$X = V_{0x}.t = 40.8 = 320m.$

Example: Look at the given pictures and find which one of the vectors given in the second figure is the relative velocity of **A** with respect to **B**.

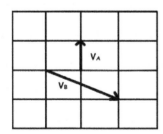 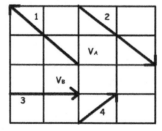

Since the observer is B, we find relative velocity of A with respect to B with following formula;

$V_{AB} = V_A - V_B$

Using vector addition properties we find relative velocity as given figure below.

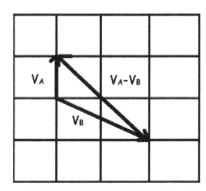

Example: Velocity of the river with respect to ground is 2m/s to the east. Width of the river is 80m. One boat starts its motion on this river at point A with a velocity shown in the figure below. Find the time of the motion and horizontal distance between the arrival point and point A.

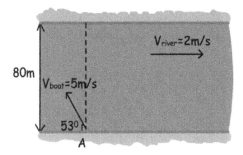

Components of boat velocity;

$V_x = 5 \cdot \cos 53^0 = -3$ m/s to the west

$V_y = 5 \cdot \sin 53^0 = 4$ m/s to the north

Time for passing the river is;

t=X/V=80m/4m/s=20s

Resultant velocity in horizontal is;

$V_R = V_x + V_{river}$

$V_R = -3 + 2 = -1$ m/s to the west

Distance taken in horizontal is;

X=V.t

X=1m/s.20s=20m

Example: A boat in a river having constant velocity travels 120m distance from point A to B in 20 s and turns back from B to A in 12 s. If the velocity of the river is zero, find the time of this trip.

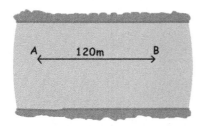

Since the time of trip from B to A is longer than the time of trip from A to B, direction of river velocity is to the west.

Velocity of river with respect to ground is V_{river}, and velocity of boat with respect to river is $V_{boatriver}$.

Velocity of boat with respect to ground when it travels from A to B becomes;

$V_b = V_{boatriver} - V_{river}$
And when it travels from B to A;

$V_b = V_{boatriver} + V_{river}$
We can find velocities using following formula;

1. $V_{boatriver} - V_{river} = 120/20 = 6 m/s$
and

2. $V_{boatriver} + V_{river} = 120/12 = 10 m/s$
Solving equations 1. and 2. we find the velocities of river and boat.

$V_{boatriver} = 8 m/s$ and $V_{river} = 2 m/s$
If the velocity of river is zero, boat travels 240m distance in;

$240 = 8 m/s \cdot t$

$t = 30 s$

Example: Two swimmers start to swim at the same time as shown in the picture below. If they meet at point C, find the ratio of their velocities with respect to water.

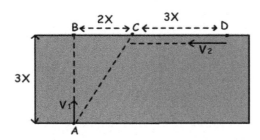

If there is no river velocity, swimmer1 reaches point B, and swimmer2 reaches point B, however, river velocity makes them reach to point C. In other words, swimmer1 having velocity V1 takes 3X distance and swimmer2 having velocity V2 takes 5X distance during same time. Thus;

$V_1/V_2 = 3/5$

DYNAMICS

In this unit we will deal with causes of motion. What makes objects move is, our primary concern. Moreover, we give Newton's law of motion and try to explain causes of motion with these laws. Let's begin with concepts one by one that will help us in analyzing motion.

FORCE

What change the state of object is called *"force"*. We mean by saying state, shape or position of the object. You can change the state of ball by kicking it or you can change the shape of spring by pressing it. In all examples you apply force to objects and make changes in their states; shapes or positions. In summary, you observe and apply force every day. Force is a vector quantity having both magnitude and direction. We understand that cause of motion is "force". The unit of force is Newton or $kg.m/s^2$.

PROJECTILE MOTION

  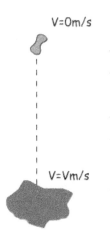

We have learned one-dimensional motion in previous topics. Now, we will deal with two-dimensional motions. In general people say that motion in two dimensions is called projectile motion. However, this definition is not enough to explain projectile motion. We can generally define it as "**motion taking place under only force of gravity**". In this type of motion gravity is the only force acting on our objects. We can observe different types of projectile motion examples in daily life. For example, you throw the ball straight upward, or you kick a ball and give it a speed at an angle to the horizontal or you just drop things and make them free fall; all these are examples of projectile motion.

In projectile motion, gravity is the only force acting on the object. I will explain this sentence with a picture and examples. First, look at the given picture, which shows the motion path, velocities at different points and forces acting upon the object doing projectile motion.

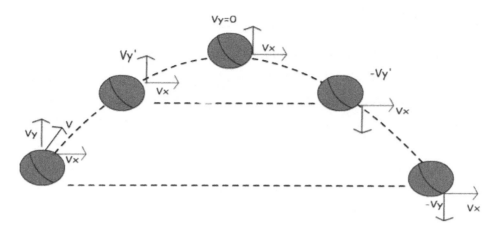

As you can see in the picture given above, we have a projectile motion and velocity components at different positions. At the beginning ball is thrown with an angle to the horizontal. V is its velocity and its direction is northeast. V_x and V_y are the X and Y components of our velocity. If we look at the forces acting on our ball we see only gravity as a force. We examine our motion in two parts, first one is horizontal motion and second one is vertical motion. When we look at the vertical motion of the object we see that it looks like example solved in free fall motion section. In vertical as you can see in the picture, our velocity is decreasing in the amount of gravitational acceleration. At the top where it reaches its maximum height vertical component of our velocity becomes zero as in the case of free fall examples. After V_y becomes zero our ball changes its direction and make free fall now. At the same levels, magnitudes of V_y are the same however their signs are opposite. Right side of our picture has "-"sign in front of the V_y because its direction is downward. Finally, when the ball hits the ground V_y reaches its beginning magnitude but opposite in direction. We see effect of gravity on vertical motion. Now, let's look at the horizontal part of our projectile motion. This part is so easy that you can understand from the picture, our horizontal component of velocity is constant during the motion. Why it is constant? What changes the velocity? In previous section we learned force concept that causes change in the state of motion. Look at our horizontal motion carefully. Is there any force acting on our object in horizontal direction +X or –X? The answer is actually no. However, in –Y direction gravity is acting on our object which makes V_y decrease and becomes zero at the top. All these explanations say that, we have two motions in projectile motion. One of them is constant motion in horizontal and other one is free fall under the effect of gravity in vertical. We tried to explain projectile motion with words. Now it is time to give equations of motion under two titles.

Vertical Motion

In vertical direction, we said that gravity acts on our objects and gives negative acceleration "-9,8m/s²" to the ball. This means that, our velocity decreases -9,8m/s² in each second. We find velocity of free falling object by equation **V=g.t**. In this situation, we have initial velocity and our equation becomes;

$V=V_it+gt$ where acceleration is $-9{,}8 m/s^2$

The distance in free fall is calculated by the equation;

$$\text{Distance Traveled} = \frac{1}{2} g \cdot t \cdot t = \frac{1}{2} g \cdot t^2$$

As in the velocity case, our distance is calculated considering initial velocity of the object by given formula;

$$\text{Distance Traveled} = V_i t - \frac{1}{2} g \cdot t^2$$

We put "-"sign because, direction of g is downward.

Horizontal Motion

We have constant motion in horizontal because; there is no force acting on our object. Thus, the X component of velocity is constant and acceleration in X direction is zero. The equation that is used to calculate distance and velocity is given below.

$X = V \cdot t$

You can find distance traveled, time elapsed and velocity of the object, if two identity are given in this equation.

Now I will solve some examples related to each type of projectile motion.

Example: In the given picture below, Alice throws the ball to +X direction with an initial velocity of 10m/s. If time elapsed during this motion is 5s, calculate the height object is thrown and Vy component of the velocity after it hits the ground.

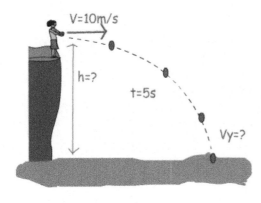

In vertical direction we have free fall motion.
$h = \frac{1}{2} g \cdot t^2 = \frac{1}{2} 10 \cdot 5^2$
$h = 125m$

$V_y = -g \cdot t$
$V_y = -10 m/s^2 \cdot 5s$
$V_y = -50 m/s$

In horizontal since our velocity is constant;
$X = V \cdot t$
$X = 10 m/s \cdot 5s = 50m$

Example: In the given picture you see the motion path of cannonball. Find the maximum height it can reach, horizontal distance it covers and total time from the given information. (The angle between cannonball and horizontal is 53° and sin53°=0, 8 and cos53°=0, 6)

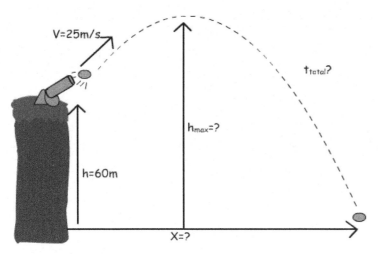

56

First, we find the components of the velocity.

$V_x = V \cdot \cos 53°$ $V_y = V \cdot \sin 53°$
$V_x = 25 \cdot 0{,}6 = 15\,m/s$ $V_y = 25 \cdot 0{,}8 = 20\,m/s$

Motion in vertical;

$V_{final} = V_{initial} - g \cdot t$
$0 = 20\,m/s - 10 \cdot t$
$t = 2s$ 2s is the time required for cannonball to reach maximum height

$h = V_y \cdot t - \frac{1}{2} g \cdot t^2$
$h = 20\,m/s \cdot 2s - \frac{1}{2} \cdot 10\,m/s^2 \cdot (2s)^2$
$h = 20m$

$h_{max} = 20m + 60m = 80m$

Free fall from the maximum height;
$h = \frac{1}{2} g \cdot t^2$
$80m = \frac{1}{2} \cdot 10\,m/s^2 \cdot t^2$
$t = 4s$
$t_{total} = 4s + 2s = 6s$

Motion in horizontal;

$X = V \cdot t$
$X = 15\,m/s \cdot 6s$
$X = 90m$

Example: John kicks the ball and ball does projectile motion with an angle of 53° to horizontal. Its initial velocity is 10 m/s, find the maximum height it can reach, horizontal displacement and total time required for this motion. ($\sin 53° = 0,8$ and $\cos 53° = 0,6$)

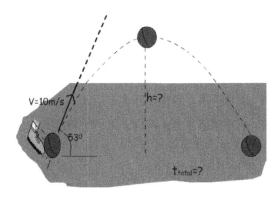

First we seperate our velocity into its components to make problem simple.

$V_x = V \cdot \cos 53° = 10 \cdot 0{,}6 = 6\,m/s$
$V_y = V \cdot \sin 53° = 10 \cdot 0{,}8 = 8\,m/s$

In horizontal;
$X = V \cdot t$
$X = 6 \cdot 1{,}6 = 9{,}6m$

In vertical;
$V = V_0 - g \cdot t$
$0 = 8 - 10 \cdot t$
$t = 0{,}8s$ 0,8s is the time required only for the half of the motion. Thus, we multiply it with two for total time.
$t_{total} = 1{,}6s$

$h = \frac{1}{2} g \cdot t^2$
$h = \frac{1}{2} \cdot 10 \cdot (0{,}8s)^2 = 3{,}2m$ We use 0.8s time because we just consider vertical motion.

NEWTON'S LAWS OF MOTION

Newton's First Law of Motion

In his first law of motion Newton stated that all objects save their states of motion. In other words, object staying at rest continuous to stay at rest, on the contrary moving objects continue to move unless a nonzero force is applied on them. What we mean by a word "nonzero force"? Look at the given picture to understand what we mean.

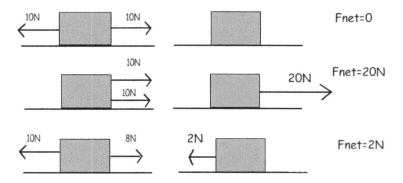

We call nonzero force as "net force" also. As you see from the picture, if forces acting on an object are in same direction they are summed, if they are in opposite directions we take one of them in negative direction and make calculations considering their signs and find resultant force vector. In first situation, applied forces are opposite direction and since their magnitudes are same the net force becomes zero. In second situation, forces are in same direction and they are also same in magnitude thus, resultant vector becomes sum of them. And in final situation, forces are in opposite directions and have different magnitudes. To find resultant force, we make vector summation. Direction of resultant vector is same with larger applied force. After these explanations, I think "net force" is clear in your mind. Now let's turn to our main topic "Newton's first law of motion" in other words "law of inertia". We said that, objects want to continue their state of motion whether they are at rest or in motion. If there is no net force on an object at rest than it continues to be at rest, if there is no net force on moving object, it continues to move at constant speed. Look at given picture below.

As you see, passengers move forward when driver breaks the bus. Before the brake passengers have same velocity with bus. Thus, because of Newton's first law, they tend to move with same speed. However, a sudden brake resulting of a net force makes passengers move forward. This is a good example for law of inertia from daily life. We can increase the number of examples. For example you all experience change in your body when in an accelerating elevator or at a swing. To sum up, keep following diagram in your mind about Newton's first law of motion.

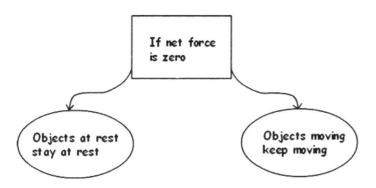

Equilibrium Conditions

In physics, equilibrium means "forces are in balance". The net force applied on an object should be zero. In other words, forces acting downward and acting upward, and forces acting right and acting left should be equal in magnitude. Look at example given below and try to understand what I say.

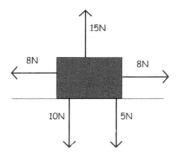

In given picture, there are five forces acting on our block. Let us analyze them, whether this block is in equilibrium or not. Forces acting downward are; **10N+5N=15N** Force acting on upward is; **15N**. Since the directions are opposite and magnitudes are equal these forces balance each other. Thus, we can say that, our block is in equilibrium in +y, -y direction. Now, look at forces acting right and left.

As you see, they are also equal in magnitude and opposite in direction. As a result, since the forces applied at four directions balance each other or net force applied on block is zero; block is in equilibrium.

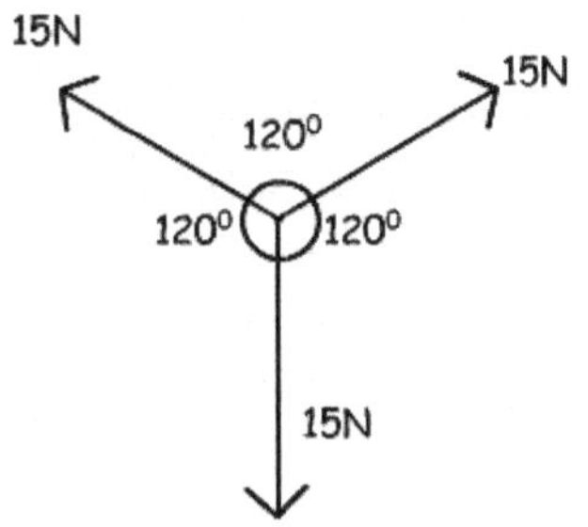

The given system is also in equilibrium. Let's prove this.

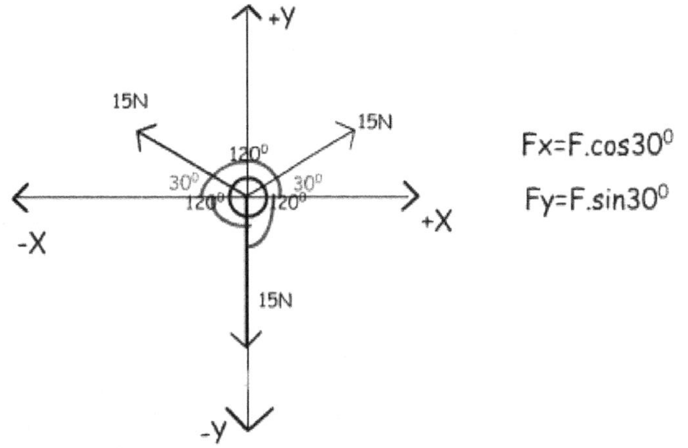

$F_x = F \cdot \cos 30°$

$F_y = F \cdot \sin 30°$

I use a little bit trigonometry to find X and Y components of given forces. As you can see from given picture, X and Y components of vectors have same magnitudes but opposite directions. (In vectors chapter, we have learned that; if there is 120° angle between two equal forces, their resultant vector is equal to one of them.) Thus, the net force acting on object becomes zero, which is the condition of equilibrium.

Example: If the boy is in equilibrium, find G from the given information in picture.

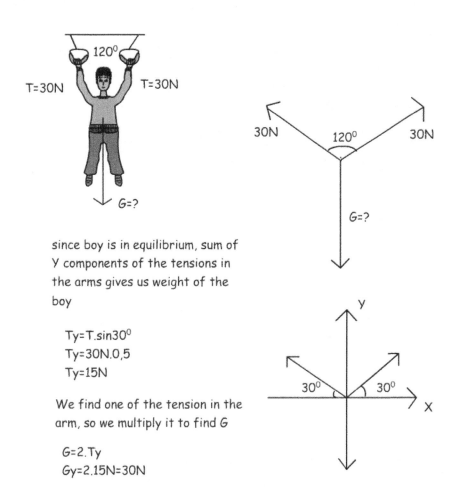

since boy is in equilibrium, sum of Y components of the tensions in the arms gives us weight of the boy

$T_y = T \cdot \sin 30°$
$T_y = 30N \cdot 0{,}5$
$T_y = 15N$

We find one of the tension in the arm, so we multiply it to find G

$G = 2 \cdot T_y$
$G_y = 2 \cdot 15N = 30N$

Since tensions in the arms are equal and angle between them is 120°, resultant forces is equal one of them. In other words, resultant force of tensions is 30 N and it is equal to weight of the boy. Knowing these tips helps you to solve problems without doing any calculations.

Newton's Second Law of Motion

In previous topics I said that force causes acceleration. Moreover, we also learned the net force concept in last section. Now, we deal with relation between force and acceleration. As you remember, acceleration is the rate of change in velocity of object. This change occurs because of the net force. Thus, we can say that there is a linear relation between the net force acting on object and acceleration. We show this relation like;

Net force ∼ acceleration

If we increase the amount of net force than acceleration also increases with the same amount. If we decrease the net force than acceleration also decreases. Let's see it from given picture below.

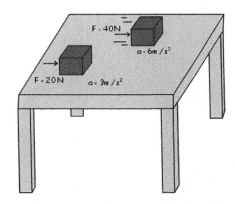

In the Picture given above, we double the force applied on box and we see that acceleration also becomes double. On the contrary we decrease the force then, acceleration also decreases. We understand the relation between force and acceleration. Now, let's talk about relation between mass and acceleration. Do you think mass affects the acceleration? Suppose that you push a box that is empty, you can easily push it. If the box is full, then can you push it easily with the same force? The answer is of course "NO". You all experience this kind of examples in your daily life. Bigger the mass results in bigger the force. Thus, we find another relation of force and mass, which is;

Net force ∼ mass

We found two relations of force. Now it is time to combine them.

$$\left. \begin{array}{l} \text{Net Force} \sim \text{Acceleration} \\ \text{Net Force} \sim \text{Mass} \end{array} \right\} \text{Net Force} \sim \text{Mass.Acceleration}$$

Force is linearly proportional to mass and acceleration. If the mass is constant, when we increase force, object gains acceleration with the same amount or, if the force is constant, when we decrease mass acceleration increases with the same amount. Beyond of these explanations, we can easily write formula of Newton's second law of motion.

F=m.a

Where, *F* is force and its unit is Newton, *m* is mass and has unit kg and *a* is acceleration has unit m/s².

Example: Find acceleration of the block given in the picture below.

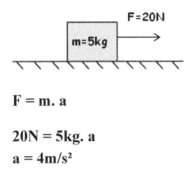

F = m. a

20N = 5kg. a

a = 4m/s²

How can we find direction of acceleration? It is a vector quantity, so it must have a direction beside its magnitude. Look at the example given above mass is a scalar quantity and force is a vector quantity. So, net force is the only quantity, which determines the direction of acceleration. For this example, our acceleration is 4m/s² to the rightward.

Example: In the picture given below, a horse is pulling the horse box having 8 kg mass in it with a force of 40N; if the applied force has an angle of 37° to horizontal. Calculate the acceleration of horse box.

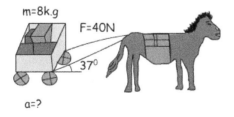

First, we should find the x and y components of force.

Fx=F.cos37°   and   Fy=F.sin37°
Fx=40N.0,8   Fy=40N.0,6
Fx=32N   Fy=24N

Since the motion is in X direction, we take the x component of the force.

F=m.a   The horsebox is
32N=8kg.a   traveling with
a=4m/s²   4m/s² acceleration.

Direction of acceleration for this example is same with direction of Fx; because Fx is the only force in direction of motion.

Mass and Weight

Mass and weight are the most confusing concepts in physics. Sometimes we use them interchangeably in daily life; however, in physics we must be careful while using them. Mass is the quantity of matter. As given in the definition it contains only the magnitude thus we say that it is a scalar quantity. Mass is constant everywhere.

Weight is of course related to the mass, but it has a little bit different definition. Weight is the force of gravity acting upon the things. Since it is a kind of force, weight is a vector quantity. Weight of the object can be different in different part of the world. Since it is linearly proportional to gravitation, it changes when the value of gravitation changes. Now let's calculate **weight** in other words **force of gravity**.

$W = m \cdot g = m \cdot 9.8 \text{ m/s}^2$
Where; m is the mass and g is the gravitational acceleration.

Example: Find weight of the object having mass 15 kg.

$W = m \cdot g = 15\text{kg} \cdot 9,8 \text{ m/s}^2$

W= 147 N

Newton's Third Law of Motion

In this law Newton states that, when we apply a force on something then it also applies a force on us with same magnitude but opposite in direction. In general, all actions have reactions in the same magnitude but opposite direction. Suppose that, when you swim you push the water to the backward direction and water also pushes you to the forward direction. We can increase the examples, when birds fly they push the air with their wings and air also push the bird in the opposite direction, in this way birds can stay in the air and fly. In everywhere, we can see this couple of forces. Every actions cause reactions. Look at given pictures, and see the action reaction force couples.

Force exerted by rifle to bullet gives acceleration to bullet, moreover, bullet also exerts a force to the rifle, which is the reaction of action force, and as a result rifle is recoiled.

Look at the picture given below. The box applies a force because of its weight, and table shows a reaction to this action. These force pairs are same in magnitude but as you see their directions are opposite. In following topic we will examine this example in detail.

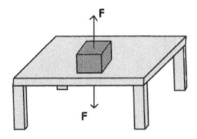

Normal Force

Contact objects (as table book example given above) exert force to each other because of their weights. In this example, book exerts a force to table because of its weight and table also exerts force to the book. We call this force as "normal force" which is same in magnitude and opposite in direction with the applied force (weight of the book). For different situations, we say that in general normal force is the reaction to the perpendicular force exerting on it. We will deal with different examples of normal force for clear understanding. Look at given examples below and follow the steps to understand how we can find normal force for different situations.

Example: Find the normal force that the inclined plane exerts on the box.

(Sin37°=0, 6 cos37°=0, 8), (m=4kg, g=10m/s²)

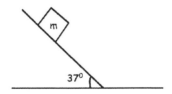

Since our box is on inclined plane we cannot say directly the normal force is equal to the weight of the object. Normal force is equal to the force exerting on it **perpendicularly**. Let's draw a free body diagram and show all the forces on box.

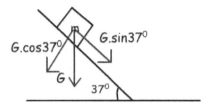

Red arrows show the force of weight and its components. Horizontal and vertical components of force are found by using trigonometry. We said that for the normal force, there must be perpendicular force acting on the contact surface. In this example, cause of normal force is the vertical component of weight. Let's calculate the value of it.

$F_{normal} = G \cdot \cos 37^0$ where, $G = m \cdot g$
$F_{normal} = m \cdot g \cdot 0,8$
$F_{normal} = 4 kg \cdot 10 m/s^2 \cdot 0,8$
$F_{normal} = 32 N$

Look at the given pictures.

In the first picture, a boy exerts a force to ball by kicking it and ball also exerts reaction force to this action force. Moreover, in the second picture again forces are shown with arrows. You should be careful while drawing these arrows. Most of the students have difficulty in this point. Be careful, forces are acted on different objects; thus, we should show them on different bodies. This explanation also answers frequently asked tricky question "If the forces are equal in magnitude and opposite in direction why they do not cancel each other". From the explanation given above we say that, they do not cancel each other because they are acting on different bodies or systems. Forces should be on the same object or system to cancel each other.

Friction Force

Friction force results from interactions of surfaces. Irregularities in the structure of the matters cause friction force. These irregularities can be detected in micro dimensions.

You may not see any irregularity on the surface of the material however it does exist. Friction force is always opposite to the direction of motion and tends to decrease net force. All materials have their own friction constant. In other words, friction force depends on the type of materials. Another factor affecting friction force is the normal force. When you apply a force to an object, then friction force becomes active and resists with the force of having opposite direction to your net force.

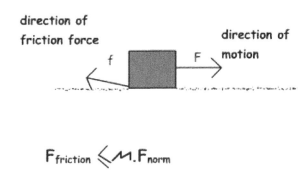

$$F_{friction} \leq \mu \cdot F_{norm}$$

We can calculate the friction force by this formula;

$$F_{friction} \leq \mu \cdot F_{norm}$$

Where, µ is the coefficient of friction and it depends on type of material. F_{norm} is the reaction of surface to the object because of its weight (Normal Force).

$$F_{friction} \leq \mu \cdot m \cdot g$$

In this equation, we assume that object is on flat surface and normal force of the surface is equal to weight of object; **m.g**. However, if the object is on an inclined plane, than we take the vertical component of weight while calculating normal force of the surface.

Friction can be studied under two topics; static friction and sliding friction. Surfaces apply different friction constant when the object is at rest and sliding. Interestingly, friction constant of the objects at rest is higher than the friction constant of sliding objects. The sliding friction force is calculated by using coefficient of friction (**µ**) and F_{norm} (normal force that surface apply to object).

$$F_{friction} = \mu_{sliding} \cdot m \cdot g \quad \text{sliding}$$

Friction force also exists when there is no motion. If two objects are in contact then we can talk about friction force; there is no need for motion to have friction. In static friction force, two objects are in contact however there is no motion in other words object does not slide on the surface. You all experience the static friction in daily life. For example, suppose that you push a huge box, which is on the carpet, however, box does not move. Static friction becomes exist when you apply a force to the object. The amount of the static friction is equal to the amount that you apply and the direction is opposite to the direction of the motion. If you increase the applied force until one point static friction also increases. We will calculate this limit point by the formula given below.

F$_{friction}$≤ μ$_{static}$.m.g   static

We use ≤ symbol instead of =, because static friction changes with changing applied force. It has the value of 1 to the limit value. We calculate limit value by the formula given above. Most of the time, magnitude of static friction is greater than magnitude of sliding friction for same surfaces. Now we solve some problems related to friction and Newton's law of motions.

Example: 50N of force is applied to the 6 kg box. If the coefficient of friction is 0, 3, find the acceleration of the box. (Sin53°=0, 8 and cos53°=0, 6)

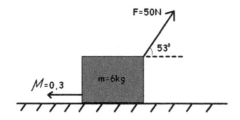

Components of force;

F$_x$ = F.cos53° = 50N. 0, 6 = 30N

F$_y$ = F. sin53° = 50N. 0, 8 = 40N

Free body diagram of the system is given below. Using this diagram we find normal force of the surface and friction force.

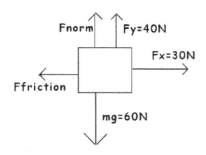

F$_{normal}$=mg-Fy

F$_{normal}$= 60N – 40N = 20N

And friction force is;

F$_{friction}$=μ.**F**$_{normal}$=0, 3.20N=6N

Net force in –Y, Y direction is zero, in other words, box is in equilibrium in this direction. However, in –X, +X direction net force is not zero so there must be a motion and acceleration in this direction. We find acceleration of the system by using Newton's second law of motion.

F$_{net}$=m.a

F$_x$-**F**$_{friction}$=m.a

30N-6N=6kg.a

a=4m/s²

Example: Find the acceleration of the system. (μ=0,4, sin37°=0,6, cos37°=0,8 and g=10m/s²)

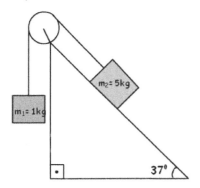

Free body diagram of the system is given below.

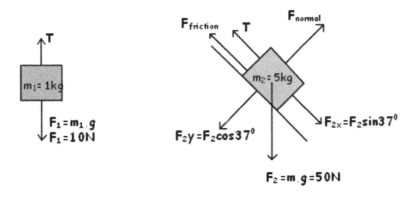

Now, let's calculate components of F$_2$ and acceleration of the system. T is equal to F$_1$ 10N.

F$_{2x}$ = F$_2$.sin37⁰ = 50.0,6 = 30 N

F$_{2y}$ = F$_2$.cos37⁰ = 50. 0, 8 = 40 N

$F_{normal} = F_{2y} = 40 \text{ N}$

$F_{friction} = \mu . F_{normal} = 0,4.40 = 16 \text{ N}$

$F_{net} = F_{2x} - F_{friction} - T = 30N - 16N - 10N = 4 \text{ N}$

$F_{net} = m.a = (1 + 5) kg.a = 4N \quad a = 0,22 \text{ m/s}^2$

MORE PROBLEMS RELATED TO DYNAMICS

Example: A box is pulled with 20N force. Mass of the box is 2kg and surface is frictionless. Find the acceleration of the box.

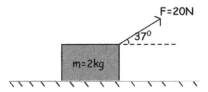

We show the forces acting on the box with following free body diagram.

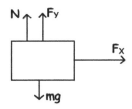

X component of force gives acceleration to the box.

$F_X = F \cdot \cos 37^0 = 20 \cdot 0,8 = 16N$

$F_X = m \cdot a$

$16N = 2kg \cdot a$

$a = 8m/s$

Example: Picture, given below, shows the motion of two boxes under the effect of applied force. Friction constant between the surfaces is k=0,4. Find the acceleration of the boxes and tension on the rope. ($g=10m/s^2$, $\sin 37^0 = 0,6$, $\cos 37^0 = 0,8$)

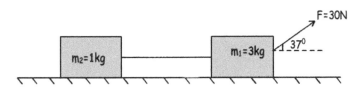

Free body diagram of these boxes given below.

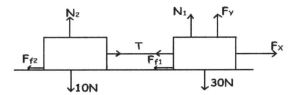

Components of force,

$F_X = F \cos 37° = 30 \cdot 0{,}8 = 24N$

$F_Y = F \sin 37° = 30 \cdot 0{,}6 = 18N$

$N_1 = m_1 \cdot g - F_y = 30 - 18 = 12N$

$N_2 = 10N$

F_{f1} and F_{f2} are the friction forces acting on boxes.

$F_{f1} = k \cdot N_1 = 0{,}4 \cdot 12 = 4{,}8N$ and $F_{f2} = k \cdot N2 = 0{,}4 \cdot 10 = 4N$

We apply Newton's second law on two boxes.

m_1: $F_{net} = m \cdot a$

$20 - T - F_{f1} = 3 \cdot a$ $20 - T - 4{,}8 = 3 \cdot a$

m_2: $T - F_{f2} = 1 \cdot a$ $T - 4 = a$

$a = 2{,}8 m/s^2$

T=6,8N

Example: you can see in the picture given below, two boxes are placed on a frictionless surface. If the acceleration of the box X is $5m/s^2$, find the acceleration of the box Y.

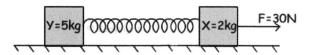

Free body diagrams of boxes are given below;

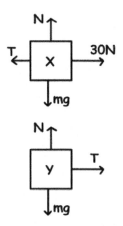

$F_{net} = m \cdot a$

$(30 - T) = 2 \cdot 5$

T=20N

F_net=m.a

T=5.a

20=5.a   a=4m/s²

Example: In the system given below ignore the friction and masses of the pulleys. If masses of X and Y are equal find the acceleration of the X? (g=10m/s²)

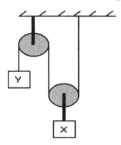

Free body diagrams of boxes are given below;

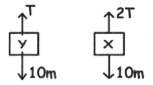

Since force acting on X is double of force acting on Y, **a_X=2a_Y**

For X: 2T-10m=m.a

For Y: T-10m=m.2a

a=2m/s²

Example: When system is in motion, find the tension on the rope.

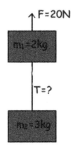

Free body diagrams of boxes are given below;

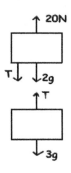

m₁: T+2g−20=2.a

m₂: 3g−T=3.a

5g−20=5.a

a=g−4 putting it into m₁ equation;

T+2g−20=2(g−4)

T=12N

Example: Net force vs. time graph of object is given below. If displacement of this object between t-2t is 75m, find the displacement of the object between 0-3t.

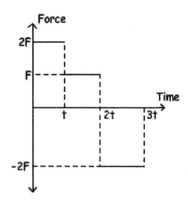

We draw acceleration vs. time graph using force vs. time graph of the object.

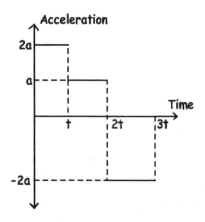

Area under the graph gives velocity.

If we say at=V then,

Vt=2V

V2t=3V

V3t=V

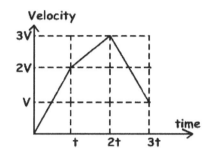

We draw velocity vs. time graph now.

Area under the velocity vs. time graph gives us displacement of the object.

0-t: ΔX_1=2Vt/2=Vt

t-2t: ΔX_2=5/2.Vt

2t-3t: ΔX_3= 2.Vt

We know ΔX_2=5/2.Vt=75m, Vt=30m

Total displacement=ΔX_1+ΔX_2+ΔX_3=Vt+5/2.Vt+2Vt

Total displacement=30+75+2.30=165m

Example: An object is pulled by constant force F from point A to C. Draw the acceleration vs. time graph of this motion. (F>mg.sinθ and surface is frictionless.)

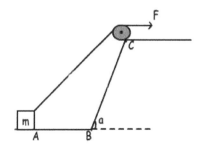

Motion of the box between points A to B:

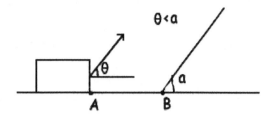

F.cosθ=m.a₁

When the object gets closer to point B, θ becomes larger, and value of cosθ decreases. Thus, a_1 decreases between the points A -B.

Motion between points B-C

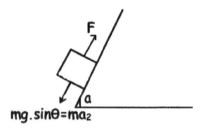

mg.sinθ=ma₂

Net force between points B and C is constant. Thus, a_2 is also constant. Acceleration vs. time graph of the box is given below;

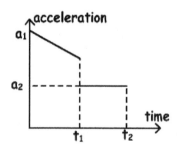

WORK POWER ENERGY

WORK

Suppose that, a force is applied on an object and object moves in the direction of applied force then we said *work* has done. Let me explain in other words. There must be a force applied to an object and object must move in the direction of the applied force. If the motion is not in the direction of force or force is applied to an object but there is no motion then we cannot talk about work. Now we formulate what we said above.

Work = Force. Distance

As you can see, work is equal to scalar multiplication of force and distance or we can say that; work is a scalar quantity. We will symbolize force as **"F"**, and distance as **"d"** in formulas and exercises. If there is an angle between force and direction of motion, then we state our formula as given below;

Work = Force. Distance. cosθ

Look at given picture. We will analyze each case in detail.

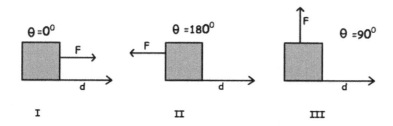

In the first case, force and distance are in same direction and angle between them is zero. Thus, cos0 is equal to 1 and work done is ; **W=F.d**

If the force and distance are in opposite directions then angle between them becomes 180 degree and cos180 is equal to -1. Work done becomes;

W = -F.d

The last case shows the third situation in which force is applied perpendicularly to the distance. Cos90 degree is zero thus, work has done is also zero. W=F.d.cos90°=0

Now let's talk about unit of work. From our formula we found it kg.m²/s² however, instead of this long unit we use *joule*. In other words;

1 joule=1N.1m

Examine the force vs. displacement graph given below. You will see that, area under this graph also gives us work has done.

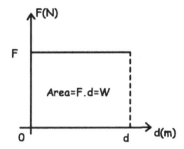

If the force applied on an object is variable, work has done can be found by using average force or using area under force vs. distance graph. Following graph shows positive and negative work.

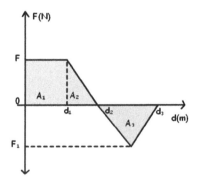

Look at the given examples below; we will try to clarify work with examples.

Example: Force of 25 N is applied to a box and box moves 10m. Find work done by this force. (Sin37°=0, 6 and cos37°=0, 8)

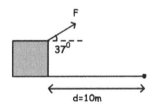

Since box moves in X direction, we should find X and Y components of the applied force. There is no motion in Y direction, we ignore Y component of the force.

$F_x = F \cdot \cos 37°$

$F_x = 25 \cdot 0, 8 = 20N$

Work done by F_x is;

$W = F_x \cdot d = 20N \cdot 10m = 200$ joule

Motion of the box is in X direction. So, we use the X component of applied force. Since the angle between X component of force and distance is zero cos0° becomes 1. I did not mention it in the solution. If it was a different value than 1 we must write it also.

Example: Look at the given picture below. There is an apple having a force applied perpendicularly on it. However, it moves 5m in X direction. Calculate the work done by this force.

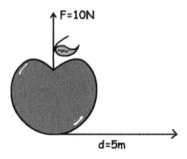

W = F.d. cos90°

W = 10N. 5m. 0 = 0 joule

Value of cos90° is zero; work done by this force becomes also zero. You can easily understand whether work is done or not by just looking at the picture. Force is applied perpendicularly to the apple however, apple moves in +X direction. Since direction of applied force and motion are different, we cannot talk about work.

Example: If box given in the picture is touching to the wall and a force is applied on it, find the work done by the force.

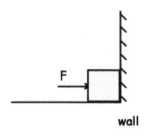

Box is touching to the wall and force cannot move it. Because there is no distance we cannot talk about work. As you can see our formula;

Work=Force. Distance If one of the variables is zero than work has done becomes zero.

POWER

Power is the rate of work done in a unit of time. It can be misunderstood by most of the students. They think that more power full machine does more work. However, power just shows us the time that the work requires. For example, two different people do same work with different times. Say one of them does the work in 5 seconds and the other does in 8 seconds. Thus, the man doing same work in 5 seconds is more power full. The shorter the time the more power full the man. Let's represent it mathematically;

$$Power = \frac{Work}{Time\ Interval}$$

The unit of power from the equation given above is; *joule/s*, however, we generally use *watt* as unit of power.

1joule/s=1watt

Example: Find power of the man who pushes the box 8m with a force of 15N in 6 seconds.

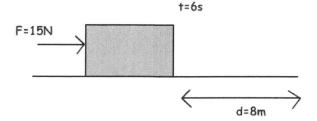

To find power, we should first find the work done by the man.

W = F.d = 15N.8m = 120 joule

Power =Work/Time = 120 j / 6s = 20 watt

Power of the man is 20 watt. In other words he does 20 joule work in 6 seconds.

BE CAREFUL Amount of power does not show the amount of work done. It just gives the time that work requires.

ENERGY

The capability of doing work is called energy. If something has energy then it can do work. It has the same unit with work *"joule"*. Energy can exist in many forms in universe. For example, potential energy in the compressed spring, kinetic energy in the moving object, electromagnetic energy and heat are some of them. In this unit we will deal with mechanical energy of the substances. Mechanical energy is the sum of potential energy and kinetic energy of the system. Let' see them one by one.

Potential Energy

Objects have energy because of their positions relative to other objects. We call this energy as *potential energy*. For example, apples on the tree, or compressed spring or a stone thrown from any height with respect to ground are examples of potential energy. In all these examples there is a potential to do work. If we release the spring it does work or if we drop the apples they do work. To move the objects or elevate them with respect to ground, we do work.

The energy of the objects due to their positions with respect to ground is called *gravitational potential* energy.

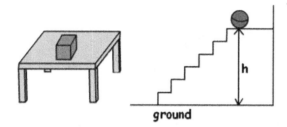

The pictures given above are the examples of gravitational potential energy. They both have a height from the ground and because of their positions they have energy or potential to do work. Look at the given example below. Spring also has potential energy because of its position.

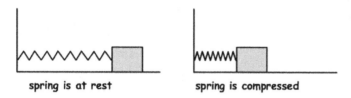

There is a spring attached between wall and box in the pictures given above. In the first picture, spring is at rest and there is no potential energy in the spring. However, in second picture, box compresses the spring and loads it with potential energy. If we release box, spring does work and pushes the box back. These two examples of gravitational and spring potential energy are calculated differently. Let me begin with calculation of *gravitational potential energy*.

First of all, we will examine factors effecting magnitude of potential energy. Work done against to the earth to elevate objects is multiplication of its *weight* and distance to the earth *(height)*. Thus, as we said before energy is the potential of doing work.

Then, gravitational potential energy becomes;

PE=weight.h =mg.h

Now, look at given picture and try to calculate potential energies of given balls for three situations.

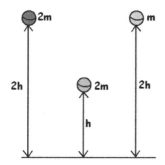

Red ball has potential energy;

PE$_1$=2m.g.2h=4mgh

Green ball has potential energy;

PE$_2$= 2m.g.h = 2mgh

Blue ball has potential energy;

PE$_3$= m.g.2h = 2mgh

Gravitational potential energy depends on weight and height of the object. Let's solve more examples to make clear it in your mind.

Example: If the potential energy of the ball in first picture is P, find potential energy of the ball in second situation in terms of P. (mass of ball is m)

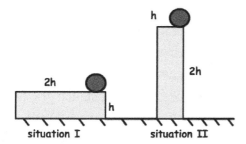

In situation I:

P = m.g.h = mgh

In situation II: (Put P in terms of mgh)

P' = m.g.2h = 2mgh = 2P

Finding Potential Energy of the Compressed or Stretched Spring: By compressing spring or stretching it you load a potential energy to it. Well, if I apply same force to different springs having different thicknesses, are they loaded with the same energy? I hear that you all say no! You are absolutely right. Of course the thinner spring is more compressed than the thicker one where the quantity of compression shows the loaded potential energy. What I want to say is that, potential energy of the spring depends on the type of spring and the amount of compression. The mathematical representation of this definition is given below.

Ep = 1/2.k.x²

Where **k** is the spring constant and **x** is the amount of compression. Picture given below shows the spring at rest.

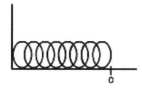

As you can see, there is no compression or stretching. Thus, we cannot talk about potential energy of the spring. However, in the pictures given below springs are not at rest position. Let's examine behavior of the springs in two situations.

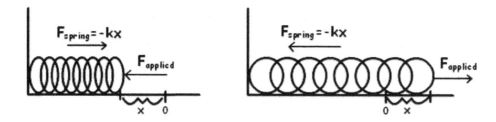

In the first picture, we apply a force, **F_{applied}**, and spring reacts this force with **F_{spring}=-kx**. The amount of compression is **x**. In the second picture, we stretch spring by amount of **x**. We apply force of F and spring gives reaction to this force with **F_{spring}=-kx** where **x** is the stretching amount and **k** is the spring constant. The given graph below is force versus distance graph of springs. We find energy equation of spring by using this graph.

As I said before area under the force vs. distance graph gives us work and, "energy" is the capability of doing work. So, area under this graph must give us the potential energy of the spring.

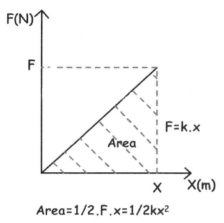

Area=1/2.F.x=1/2kx²

Example: 50N force is applied to a spring having 150N/m spring constant. Find the amount of compression of the spring.

F$_{spring}$=-kx=F$_{applied}$

50N=-150.x

X=-3m "-"shows the direction of compression.

Kinetic Energy

Objects have energy because of their motion; this energy is called **kinetic energy**. Kinetic energy of the objects having mass m and velocity v can be calculated with the formula given below;

Ek=1/2mv²

As you see from the formula, kinetic energy of the objects is only affected by the mass and velocity of the objects. The unit of the Ek is again from the formula kg.m²/s² or in general use *joule*.

Example: Find the kinetic energy of the ball having mass 0,5 kg and velocity 10m/s.

Ek=1/2mv²

Ek=1/2.0, 5. (10) ²

Ek=25joule

Example: Find the Kinetic Energy of the object at 14m from the given graph below.

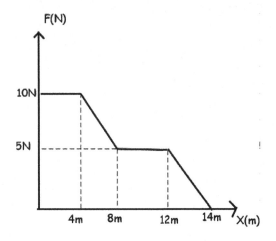

We can find the total kinetic energy of the object after 14m from the graph; we use area under it to find energy.

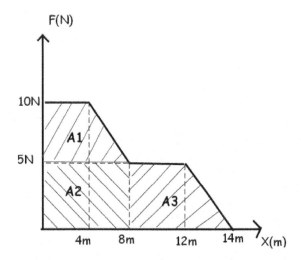

$A1 = \dfrac{(8+4) \cdot 5}{2} = 30$ $A3 = \dfrac{(6+4) \cdot 5}{2} = 25$

$A2 = 5 \cdot 8 = 40$

Total Area = A1 + A2 + A3

Total Area = 30 + 40 + 25 = 95

Ek = Total Area = 95 joule

Example: Look at the picture given below. If the final velocity of the box is 4m/s find the work done by friction.

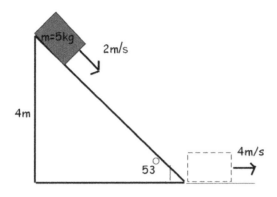

At the top;

E$_{total}$=mgh+1/2mv²

E$_{total}$=5kg.10m/s².4m+1/2.5kg. (2m/s) ²=210joule

At the bottom;

E$_{total}$=1/2mv²=1/2.5kg.(4m/s) ²

E$_{total}$=40joule

The friction uses the difference between the initial and final energy.

Work done by the friction=E$_{final}$-E$_{initial}$=210joule-40joule=170joule

Conservation Of Energy Theorem

Nothing can be destroyed or created in the universe like energy. Suppose that a ball falls from height of 2m, it has only potential energy at the beginning, however, as it falls it gains kinetic energy and its velocity increases. When it hits the ground it has only kinetic energy. Well, where is the potential energy that it has at the beginning? It is totally converted to the kinetic energy, as said in the first sentence nothing can be destroyed or created they just change form. Thus, our potential energy also changes its forms from potential to the kinetic energy. In summary, energy of the system is always constant, they can change their forms but amount of total energy does not change.

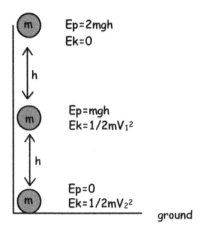

Picture shows the energy change of the ball. It has only potential energy 2mgh at the beginning. When it starts to lose height it gains velocity in other word decreasing in the amount of potential energy increases the amount of kinetic energy. At h height it has both potential and kinetic energy and when it hits the ground the potential energy becomes zero and kinetic energy has its maximum value.

$E_{initial} = E_{final}$

Example: By using the given information in picture below, find the velocity of the ball at point D.

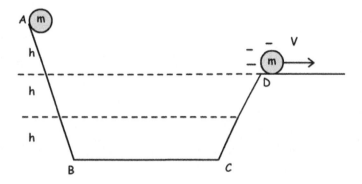

$E_i = E_f$ (conservation of energy)

$mg3h = mg2h + 1/2mV^2$

$mgh = 1/2mV^2$

$V = \sqrt{2gh}$

Example: A block having mass 2kg and velocity 2m/s slide on the inclined plane. If the horizontal surface has friction constant μ=0,4 find the distance it travels in horizontal before it stops.

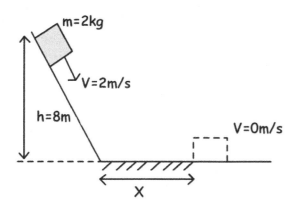

We use conservation of energy in solution of this problem.

Einitial=Efinal

Einitial =Ep+Ek=mgh+1/2mv²   **Efinal =0**

Einitial =2kg.10m/s².8m+1/2.2kg. (2m/s) ²   **Work done by friction= Einitial**

E$_{initial}$ =164joule

W$_{friction}$=μ.N.X=0,4.2kg.10m/s².X=Ei

8. X=164joule   X=20,5m

Block slides down 20,5m in horizontal.

Example: Find the final velocity of the box from the given picture.

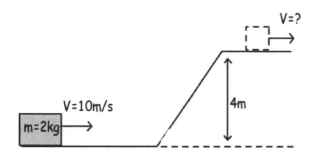

We again use the conservation of energy theorem.

E$_{initial}$ must be equal to the E$_{final}$.

$E_{initial} = E_k = 1/2 mv^2$

$E_i = 1/2 . 2kg . (10 m/s)^2 = 100 joule$

$E_{final} = E_k + E_p = 1/2 mv'^2 + mgh$

$E_{final} = 1/2 . 2kg . v'^2 + 2kg . 10 m/s^2 . 4m = 80 + v'^2$

$100 = 80 + v'^2 \quad v' = 2\sqrt{5} m/s$

Example: Find the amount of compression of the spring if the ball does free fall from 4m and compresses the spring.

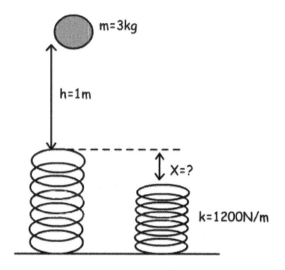

From the conservation of energy law we can find the amount of spring's compression.

$E_p = 1/2 . kx^2$ for spring

$E_{initial} = E_{final}$

$mgh = 1/2 . kx^2$

$3kg . 10 m/s^2 (1m + X) = 1/2 . 1200 N/m . X^2$

$20X^2 - X - 1 = 0$

$X = 1/4 m$ Ball compresses the spring 1/4m.

MORE PROBLEMS RELATED TO WORK POWER ENERGY

Example: In the picture given below F pulls a box having 4kg mass from point A to B. If the friction constant between surface and box is 0,3; find the work done by F, work done by friction force and work done by resultant force.

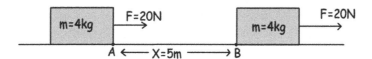

Work done by F;

$W_F = F.X = 20.5 = 100$ joule
Work done by friction force;

$W_{friction} = -F_f.X = -k.mg.X = -0,3.4.10.5 = -60$ joule
Work done by resultant force;

$W_{net} = F_{net}.X = (F-F_f).X = (20-0,3.4.10)5$

$W_{net} = 40$ joule

Example: Applied force vs. position graph of an object is given below. Find the work done by the forces on the object.

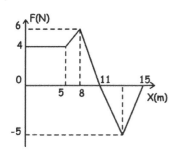

Area under the graph gives us work done by the force.

Work done between 0-5m:

$W_1 = 4.5 = 20$ joule
Work done between 5-8m:

$W_2 = (6+4)/2.3 = 15$ joule
Work done between 8-11m:

$W_3= 6.3/2=9$ joule

Work done between 11-15m:

$W_4=-5.4/2=-10$ joule

$W_{net}=W_1+W_2+W_3+W_4=20+15+9+(-10)$

$W_{net}=34$ joule

Example: Different forces are applied to three objects having equal masses. Forces pull objects to height h. Find the work done by the forces on objects and work done on gravity.

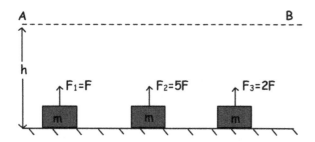

$W_1=F.h$, $W_2=5F.h$, $W_3=2F.h$

Since masses of the objects are equal, and distance taken by the objects are equal, work done on gravity of three objects are equal.

Example: In the picture given below, forces act on objects. Works done on objects by F_1, F_2 and F_3 during time t are W_1, W_2 and W_3. Find the relation of the works.

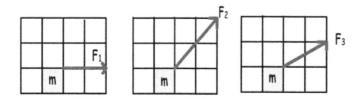

Horizontal components of the applied forces are equal to each other. Masses of the objects are also equal. Thus, acceleration of the objects and distances taken are also equal.

Work done:

$W=F_X.X$

$W_1=W_2=W_3$

Example: Applied force vs. position graph of an object is given below. Find the kinetic energy gained by the object at distance 12m.

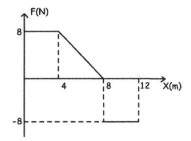

By using work and energy theorem we say that; area under the graph gives us work done by the force.

ΔEK=W=area under the graph=(8+4)/2.8-8(12-8)

ΔEK=12.4-8.4=16 joule

Example: Box having mass 3kg thrown with an initial velocity 10 m/s on an inclined plane. If the box passes from the point B with 4m/s velocity, find the work done by friction force.

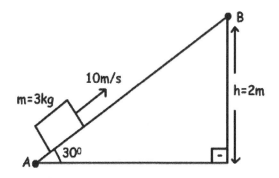

We use conservation of energy theorem.

E$_A$=E$_B$+W$_{friction}$

W$_{friction}$=1/2.m.V^2-(mgh+1/2mV$_L^2$)

W$_{friction}$=1/2.3.10^2-(3.10.2+1/2.3.4^2)

W$_{friction}$=66 joule

Example: Three different forces are applied to a box in different intervals. Graph, given below, shows kinetic energy gained by the box in three intervals. Find the relation between applied forces.

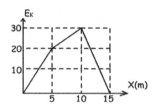

Slope of the E_K vs. position graph gives applied force

I. interval: $F_1=(20-0)/(5-0)=4N$

II. interval: $F_2=(30-20)/(10-5)=2N$

III. interval: $F_3=(0-30)/(15-10)=-6N$

$F_{III}>F_I>F_{II}$

Example: A stationary object at t=0, has an acceleration vs. time graph given below. If object has kinetic energy E at t=t, find the kinetic energy of the object at t=2t in terms of E.

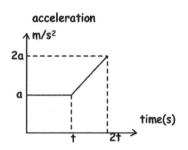

Object has velocity at t=t;

$V_1=at$

Object has velocity at t=2t;

$V_2=at+((2a+a)/2).t=at+3/2.at=5/2.at$

$V_2=5/2.V_1$

$E_2/E=(1/2.m.V_2^2)/(1/2.m.V_1^2)=(5/2.V_1)^2/V_1^2$

$E_2/E=25/4$

$E_2=25E/4$

Example: An object does free fall. Picture given below shows this motion. Find the ratio of kinetic energy at point C to total mechanical energy of the object.

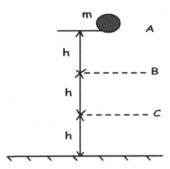

Object lost 2mgh potential energy from point A to C. According to conservation of energy theorem, this lost potential energy converted to the kinetic energy. Thus; we can say that kinetic energy of the object at point C is;

E_K=2mgh

Total mechanical energy;

E_{total}=3mgh

E_K/E_{total}=2mgh/3mgh = 2/3

Example: A box is released from point A and it passes from point D with a velocity V. Works done by the gravity are W_1 between AB, W_2 between BC and W_3 between CD. Find the relation between them.

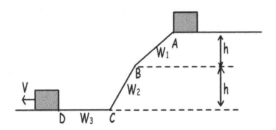

Work done by gravity is equal to change in potential energy of the object.

Interval AB: $W_1=\Delta E_p$=-mgh

Interval BC: $W_2=\Delta E_p$=-mgh

Interval CD: $W_3=\Delta E_p$=0

$W_1=W_2>W_3$

Example: An object is thrown with an initial velocity V from point A. It reaches point B and turns back to point A and stops. Find the relation between kinetic energy of object has at point A and energy lost on friction.

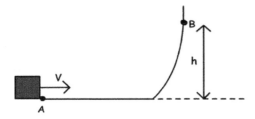

Object has kinetic energy at point A;

$E_K = 1/2 \cdot mV^2$

Object stops at point A, which means that all energy is lost on friction.

$E_K = E_{friction}$

Example: 3 rectangular plates are hanged as shown in the figure given below. If the masses of the plates are equal, find the relation between the potential energies of the plates.

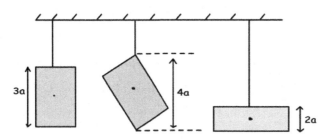

Masses of the plates are equal but center of masses are different.

$E_{P1} = m \cdot g \cdot 3a/2$

$E_{P2} = m \cdot g \cdot 2a$

$E_{P3} = m \cdot g \cdot a$

$E_{P2} > E_{P1} > E_{P3}$

Example: System given below is in equilibrium. If the potential energies of objects A and B are equal, find the mass of object A in terms of G. (Rod is homogeneous and weight of it is G.)

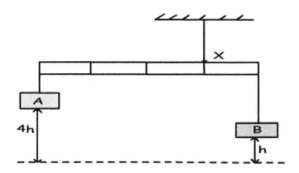

Since rod is homogeneous we can take weight of it at the center.

Equal potential energies;

$G_A \cdot 4h = G_B \cdot h$

$G_B = 4G_A$.

Moment of the system;

$G_A \cdot 3 + G \cdot 1 = G_B \cdot 1$

$3G_A + G = 4G_A$

$G_A = G$

IMPULSE AND MOMENTUM

MOMENTUM

Look at the pictures given below. If both the car and the truck have same speed, which one can be stopped first?

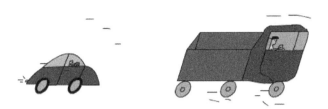

Of course all you say, it is hard to stop truck relative to car. Well, what is the reason making car stop easier? They have same speed but different masses. Can mass effect the stopping time or distance? The answer is again YES! It is hard to stop heavier objects. What we are talking about so far is **momentum**. Momentum is a physical concept that is defined as "moving body". In other words for talking about momentum we must have moving object, it must have both mass and velocity. Let me formulate what we said.

Momentum=Mass X Velocity

We show momentum in physics with "p", mass with "m" and velocity with "v". Then equation becomes:

p=m.v

Since velocity is a vector quantity and multiplied with mass (scalar quantity) momentum becomes also vector quantity. It has both magnitude and direction. Direction of momentum is the same as velocity. From the definition and given equation we can change momentum by changing its mass or changing its velocity.

Unit of the momentum is kg.m/s as you can guess from the equation.

Example: Calculate the momentum of the give objects.

A basketball ball having 2kg mass and 6m/s velocity moves to the east

A car having 15m/s velocity and 1500kg mass moves to the north

A child having mass 25kg and velocity 2m/s moves to the west

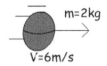
p=m.v
p=2kg.6m/s
p=12kg.m/s east

p=m.v
p=1500kg.15m/s
p=22500kg.m/s north

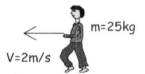
p=m.v
p=25kg.2m/s
p=50kg.m/s west

IMPULSE

We discussed above the factors changing momentum, which are mass and velocity. In most of the case mass is constant and for momentum change velocity changes. If velocity changes then acceleration occurs. In the first unit we said that force causes acceleration in other words change in the velocity is the result of applied net force. Change in the velocity is proportional the applied net force. If it is big then change is also big. Another important thing is the time of applied force. How long does it act on an object? It is linearly proportional to the change in velocity. If you apply a force on an object 1 s then you see small change in the momentum. However, if you apply force on an object long period of time then you see the amount of change in momentum is bigger than the first situation. In summary, I try to say that *impulse* is the multiplication of applied force and time interval it applied. Impulse is also a vector quantity having both magnitude and direction. It has the same direction with applied net force.

Impulse=Force.Time Interval

Impulse and momentum are directly related to each other. Let's find this relation now.

$F = m.a$

$F = m \cdot \dfrac{\text{Change in Velocity}}{\text{time interval}}$

$\underbrace{F \cdot \Delta t}_{\text{impulse}} = \underbrace{m \cdot \Delta v}_{\text{change in momentum}}$

As you can see, we found that impulse is equal to the change in momentum. In examples we will benefit from this equation.

Impulse=Change in Momentum

Example: If the time of force application is 5s find the impulse of the box given below.

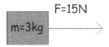

Impulse=Force.Time Interval

Impulse=15N.5s

Impulse=75N.s

Example: Find applied force that makes 10m/s change in the velocity of the box in 5s if the mass of the box is 4kg.

Impulse=Change in momentum

F.t=Δp   F.t=m. ΔV

F.t=4kg.10m/s=40kg.m/s   Impulse of the box is 40kgm/s

F=40kg.m/s/5s=8N   Applied force

All of us have chance to experience results of impulse and momentum in daily life. For example, in collisions like car crash or any other collision we can calculate the affect of force by controlling the time. Assume that, you push a box with a force of 10N for 2 seconds the impulse is 20N.s. Then you push it with a 5N force for 4 seconds and impulse does not change. As you see increasing the time decreases the amount of force. Thus, increasing the time of force application can eliminate unwanted results of force. On the contrary if you want big force then you should decrease the time and you get big force.

In this unit we will again benefit from the graphs. Look at the given graph below that shows the relationship of the force and time of a given system.

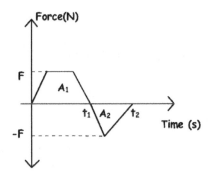

Since impulse is equal to the multiplication of force and time then, area under this graph also gives us impulse. As shown in the graph, A1 is positive impulse and A2 is negative impulse. Total impulse gives us the change in momentum as we said before. We can also draw momentum versus time and velocity versus time graph of the system.

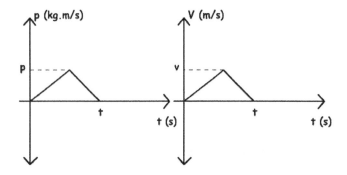

As you can see momentum vs. time graph and velocity vs. time graphs of the system are similar because momentum is directly proportional to the velocity.

Example: The graph given below belongs to an object having mass 2kg and velocity 10m/s. It moves on a horizontal surface. If a force is applied to this object between (1-7) seconds find the velocity of the object at 7th second.

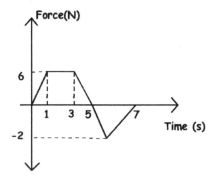

Area under the graph gives us impulse. First, we find the total impulse with the help of graph given above then total impulse gives us the momentum change. Finally, we find the final velocity of the object from the momentum change.

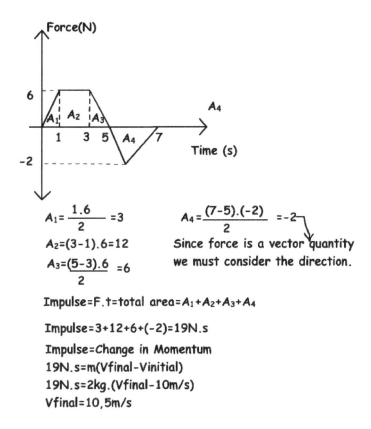

$A_1 = \dfrac{1.6}{2} = 3$

$A_2 = (3-1) \cdot 6 = 12$

$A_3 = \dfrac{(5-3) \cdot 6}{2} = 6$

$A_4 = \dfrac{(7-5) \cdot (-2)}{2} = -2$

Since force is a vector quantity we must consider the direction.

Impulse = F.t = total area = $A_1 + A_2 + A_3 + A_4$

Impulse = 3+12+6+(-2) = 19 N.s

Impulse = Change in Momentum

19 N.s = m(Vfinal - Vinitial)

19 N.s = 2kg.(Vfinal - 10m/s)

Vfinal = 10,5 m/s

CONSERVATION OF MOMENTUM

As I said before to give something acceleration we must apply an external force. If there is no force then object continue its motion. For momentum change we must apply impulse, in other words there must be external applied force to change momentum of the object. If there is no force applied then momentum of the system is conserved in magnitude and direction.

$P_{initial} = P_{final}$

To understand conservation of momentum we will examine a collision of two objects. Look at the given picture, two ball having masses m_1 and m_2 and velocities V_1 and V_2 collide. If there is no external force acting on the system, momentum of the system is conserved.

During the collision balls exert force to each other. From the Newton's third law these forces are equal in magnitude and opposite in direction. We can say action and reaction to these forces. Picture given below shows these forces at collision time.

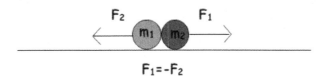

$F_1 = -F_2$

Ball having m_2 mass exerts force on m_1 and ball having m_1 mass exerts force on m_2. Since collision occurs from the interaction of the bodies then time of collision or interaction must be equal.

$F_1 = -F_2$ and $t_1 = t_2$

Impulses of the balls are;

Impulse1 = $F_1 \cdot t_1$ and   **Impulse2 = $-F_2 \cdot t_2$**

$F_1 \cdot t_1 = -F_2 \cdot t_2$

Impulses of the balls are equal in magnitude. As we said before impulse is equal to the change of momentum. Thus, we can say that momentum changes of the balls are also equal in magnitude and opposite in direction.

$m_1 (V_{1final} - V_{1initial}) = -m_2 (V_{2final} - V_{2initial})$

Conservation of momentum law says that if one of the objects loses momentum and other one gains it. Total momentum of the system is conserved.

Example: Bullet shown in the picture collides to a fixed block. 0,2 s is the interaction time of bullet with block. If the velocity of the bullet is 250m/s after the collision, find the resistance of the block to the bullet.

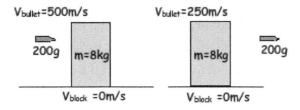

Impulse=Momentum Change

Initial Momentum=Final Momentum

pi=p_bullet+p_block
pi=0,2kg.500m/s+8kg.0m/s
pi=100kg.m/s

pf=p_bullet+p_block
pf=0,2kg.250m/s+8kg.0m/s
pf=50kg.m/s

Impulse= △p
F.0,2s=(50kg-100kg)
F=-50kg/0,2s=-250N

Resistance of the block to the bullet is -250N. We use conservation of momentum to find the change in momentum and using the impulse momentum equation we find force that block apply to bullet.

Example: Two cars are stationary at the beginning. If the car having 10kg mass starts to move to the east with a velocity of 5m/s, find the velocity of the car having mass 4kg with respect to the ground.

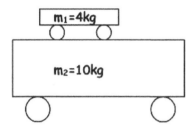

We use conservation of momentum law.
P_initial=0 since te system is stationary
P_final=p_1+p_2
p_final=m_1.v_1+m_2.v_2
p_final=4kg.v_1+10kg.5m/s
P_i=P_f
0=4kg.v_1+50kg.m/s
v_1=12,5m/s

V_1 is 12,5m/s to the west. If you want to find the velocity of the small car with respect to big car you should benefit from relative motion and do your calculations according to the vector properties, in other words you must consider the directions of the velocities.

Example: Two blocks move with the given velocities. When they collide they stick and move together. Find the velocity of the blocks after collision.

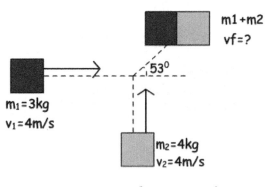

From conservation of momentum law;
$P_{initial} = P_{final}$
$P_1 + P_2 = P_f$
$P_1 = m_1 \cdot v_1 = 3kg \cdot 4m/s = 12 kg \cdot m/s$ → $P_2 = m_2 \cdot v_2 = 4kg \cdot 4m/s = 16 kg \cdot m/s$ ↑

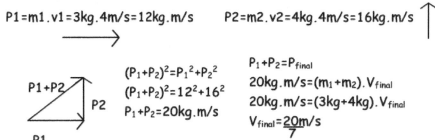

$(P_1+P_2)^2 = P_1^2 + P_2^2$
$(P_1+P_2)^2 = 12^2 + 16^2$
$P_1+P_2 = 20 kg \cdot m/s$

$P_1 + P_2 = P_{final}$
$20 kg \cdot m/s = (m_1+m_2) \cdot V_{final}$
$20 kg \cdot m/s = (3kg+4kg) \cdot V_{final}$
$V_{final} = \dfrac{20}{7} m/s$

Example: Two bullets having velocities 550m/s and 200m/s move towards to a block having mass 12kg. If the bullets stick to the block and they move together find the velocity of the final system.

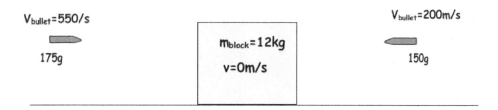

$P_{initial} = P_{final}$
$P_{bullet1} + P_{bullet2} = P_{system}$
$P_{bullet1} = m_{bullet1} \cdot V_{bullet1} = 0{,}175 kg \cdot 550 m/s = 96{,}25 kg \cdot m/s$ to the east
$P_{bullet2} = m_{bullet2} \cdot V_{bullet2} = 0{,}2 kg \cdot 150 m/s = 30 kg \cdot m/s$ to the west
$P_{initial} = P_{bullet1} + P_{bullet2} = 96{,}25 kg \cdot m/s - 30 kg \cdot m/s = 66{,}25 kg \cdot m/s$
$P_{final} = (m_{bullet1} + m_{bullet2} + m_{block}) \cdot V_{sytsem} = (0{,}175 kg + 0{,}2 kg + 12 kg) \cdot V_{sytsem}$
$66{,}25 kg \cdot m/s = 12{,}375 kg \cdot V_{sytsem}$
$V_{sytsem} = 5{,}35 m/s$ to the east

We find the momentum of the bullets first then; we calculate total initial momentum by adding them using their vector directions. Finally using conservation of momentum law we find the velocity of the system.

COLLISIONS

Momentum is conserved in all collisions. However, we can examine collisions under two titles if we consider conservation of energy. For example, if the objects collide and momentum and kinetic energy of the objects are conserved than we call this collision "elastic collision". On the other hand if the momentum of the object is conserved but kinetic energy is not conserved than we call this type of collision "inelastic *collision*". In elastic collisions some of the kinetic energy is converted to heat and sound.

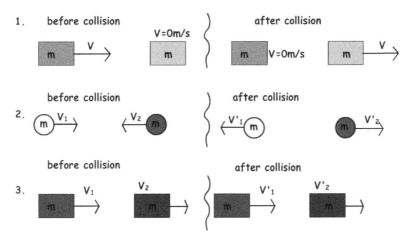

The picture given above shows the examples of elastic collision in which both kinetic energy and momentum of the system are conserved.

In this picture, which is an example of inelastic collision, momentum of the objects is conserved however; kinetic energy of the objects is not conserved.

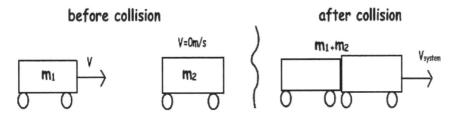

To sum up, we can say that, momentum of the system is conserved in both elastic and inelastic collisions however; kinetic energy is conserved only in the elastic collisions.

Example: A bullet that has velocity 150m/s and mass 4kg sticks to the stationary block. They move together after the collision. Find the height they reach after the collision.

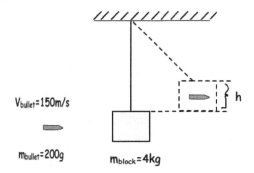

From conservation of momentum
Pi=Pf
Pi=$m_{bullet}.V_{bullet}$ Pf=$(m_{bullet}+m_{block}).V_{system}$
Pi=0,2kg.150m/s=30kg.m/s   Pf=(0,2kg+4kg).V_{system}
30kg.m/s=4,2kg.V_{system}
V_{system}=7,14m/s
From conservation of energy
Ei=Ef
1/2.m.V_{system}^2=m_{total}.g.h
1/2.0,2kg.(7,14m/s)^2=(4kg+0,2kg).10m/s^2.h
h=0,64m

Example: Look at the given picture below. Particle having mass 4m and velocity 3v explodes and breaks into two pieces. One of the pieces has mass 3m and velocity 2v. Find the momentum of the second particle and show it in a given diagram.

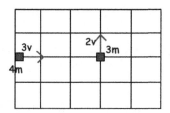

We use conservation of momentum law
Pi=Pf
Since the mass of the particle is 3m after
the explosion, second particle must have mass
m for having 4m in total.

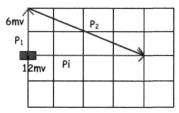

Since momentum is a vector quantity
we must consider the directions of
momentum. Thus, we find the
momentum of second particle by vector
addition method. $P_1+P_2=P_i$

The arrow shown in blue is the momentum of the
second particle.

108

MORE EXAMPLES RELATED TO IMPULSE MOMENTUM

Example: An object travels with a velocity 4m/s to the east. Then, its direction of motion and magnitude of velocity are changed. Picture given below shows the directions and magnitudes

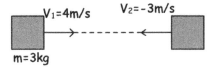

of velocities. Find the impulse given to this object.

I=F.Δt=Δp=m.ΔV
where ΔV=V$_2$-V$_1$=-3-4=-7m/s

I=m.ΔV=3.(-7)=-21kg.m/s

Example: Find the impulse and force, which make 12m/s change in the velocity of object having 16kg mass in 4 s.

F.Δt=ΔP=m.ΔV

F.4s=16kg.12m/s

F=48N

F.Δt=Impulse=48N.4s=192kg.m/s

Example: Applied force vs. time graph of object is given below. Find the impulse of the

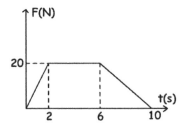

object between 0-10s.

Area under the force vs. time graph gives us impulse.

F.Δt=20.2/2+20.(6-2)+20.(10-6)/2

F.Δt=140kg.m/s

Example: Objects shown in the figure collide and stick and move together. Find final velocity objects.

$m_1 = 3$ kg, $V_1 = 8$ m/s

$m_2 = 4$ kg, $V_2 = 10$ m/s

Using conservation of momentum law;

$m_1 \cdot V_1 + m_2 \cdot V_2 = (m_1 + m_2) \cdot V_{final}$

$3.8 + 4.10 = 7 \cdot V_{final}$

$64 = 7 \cdot V_{final}$

$V_{final} = 9,14$ m/s

Example: 2kg and 3kg objects slide together, and then they break apart. If the final velocity of m_2 is 10 m/s, find the velocity of object m_1 and total change in the kinetic energies of the objects.

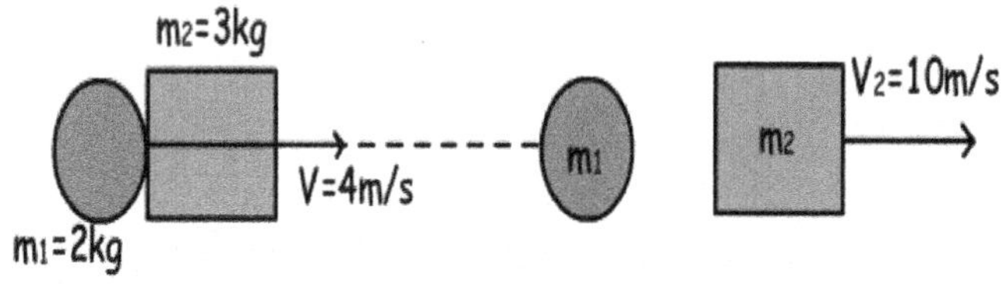

$m_1 = 2$ kg, $m_2 = 3$ kg, $V = 4$ m/s, $V_2 = 10$ m/s

Using conservation of momentum law;

$(m_1 + m_2) \cdot V = m_1 \cdot V_1 + m_2 \cdot V_2$

$5.4 = 30 + 2 \cdot V_1$

$V_1 = -5$ m/s

$E_{Kinitial} = 1/2 (m_1 + m_2) \cdot V^2$

$E_{Kinitial} = 1/2 \cdot 5 \cdot 16 = 40$ joule

$E_{Kfinal} = 1/2 \cdot 2 \cdot 5^2 + 1/2 \cdot 3 \cdot 10^2$

$E_{Kfinal} = 175$ **joule**

Change in the kinetic energy is $= 175 - 40 = 135$ joule

Example: As shown in the figure below, object m_1 collide stationary object m_2. Find the magnitudes of velocities of the objects after collision. (Elastic collision)

In elastic collisions we find velocities of objects after collision with following formulas;

$V_1' = (m_1-m_2)/(m_1+m_2) \cdot V_1$

$V_2' = (2m_1/m_1+m_2) \cdot V_1$

$m_1=6kg$, $m_2=4kg$, $V_1=10m/s$

$V_1'=(6-4/6+4) \cdot 10 = 2m/s$

$V_2'=(2 \cdot 6/6+4) \cdot 10 = 12m/s$

Example: Momentum vs. time graph of object is given below. Find forces applied on object for each interval.

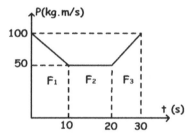

F.Δt=ΔP

F=ΔP/Δt

Slope of the graph gives us applied force.

I. Interval:

$F_1=P_2-P_1/10-0=-50/10=-5N$

II. Interval:

$F_2=50-50/10=0$

III. Interval:

$F_3=100-50/10=5N$

Example: A box having mass 0,5kg is placed in front of a 20 cm compressed spring. When the spring released, box having mass m_1, collide box having mass m_2 and they move together. Find the velocity of boxes.

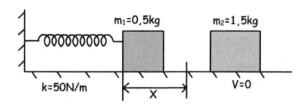

Energy stored in the spring is transferred to the object m_1.

$1/2.k.X^2 = 1/2.mV^2$

$50N/m.(0,2)^2 = 0,5.V^2$

V=2m/s

Two objects do inelastic collision.

$m_1.V_1 = (m_1+m_2).V_{final}$

$0,5.2 = 2.V_{final}$

$V_{final} = 0,5 m/s$

ROTATIONAL MOTION

Rotational motion or we can say circular motion can be analyzed in the same way of 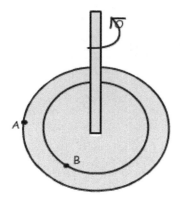 linear motion. In this unit we will examine the motion of the objects having circular motion. For example, we will find the velocity, acceleration and other concepts related to the circular motion in this section. Some concepts will be covered in this unit; rotational speed (angular speed), tangential speed (linear speed), frequency, period, rotational inertia of the objects, torque, angular momentum and its conservation.

LINEAR SPEED (TANGENTIAL SPEED)

Linear speed and tangential speed gives the same meaning for circular motion. In one dimension motion we define speed as the distance taken in a unit of time. In this case we use again same definition. However, in this case the direction of motion is always tangent to the path of the object. Thus, it can also be called as tangential speed, distance taken in a given time. Look at the given picture and try to sequence the velocities of the points larger to smaller.

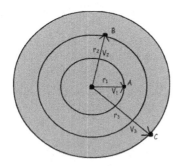

In a given period of time all points on this rotating object have same revolutions. In other words, if A completes one revolution, then B and C also have one revolution in a same time. The formula of the speed in linear motion is;

Speed=distance/time

As I said before, speed in circular motion is also defined as the distance taken in a given time. Thus, speeds of the points given in the picture below are;

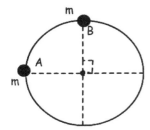

VA=Distance/time If the object has one complete revolution then distance traveled becomes; **2πr** which is the circumference of the circle object.

VA=2πr/time

Period: Time passing for one revolution is called *period*. The unit of period is **second**. **T** is the representation of period.

The equation of tangential speed becomes;

VA=2πr/T

Frequency: Number of revolutions per one second. The unit of frequency is **1/second**. We show frequency with letter **f**.

The relation of **f** and **T** is;

f=1/T

Now, with the help of the information given above lets' sequence the velocities of the points on given picture.

Since the velocity or speed of the points on rotating object is linearly proportional to the radius $r_3>r_2>r_1$;

$V_3>V_2>V_1$

To sum up, we can say that tangential speed of the object is linearly proportional to the distance from the center. Increase in the distance results in the increase in the amount of speed. As we move to the center speed decreases, and at the center speed becomes zero. We use the same unit for tangential speed as linear motion, which is "m/s".

Example: A particle having mass m travels from point A to B in a circular path having radius R in 4 seconds. Find the period of this particle.

Particle travels one fourth of the circle in 4 seconds. Period is the time necessary for one revolution. So,

T/4=4s

T=16s.

Example: If the particle having mass m travels from point A to B in 4 seconds find the tangential velocity of that particle in the given picture below. (π=3)

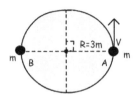

We first find the period of the motion. If the particle travels half of the circle in 4 seconds;

T/2=4s

T=8s

v=2 πR/T

v=2.3.3m/8s=9/4 m/s tangential speed of the particle

ANGULAR SPEED

Look at the given picture. If the platform does one rotation then points A and B also does one rotation. We define angular velocity as "change of the angular displacement in a unit of time". One total rotation corresponds to 2π radians. Units of angular speed are revolution per unit time radians per second. We show angular speed with the Greek letter "ω" omega. All points on the platform have same angular velocity.

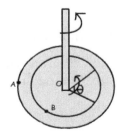

Average Speed=2π/T

ω=2π/T=2πf

Unlike tangential speed, angular speed of all points on the platform doing circular motion are equal to each other since the number of rotations per unit time are equal.

Example: If the stone does 6 rotations in 1 second find the angular velocity of it.

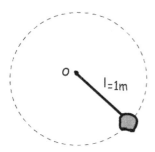

If the stone does four rotations in one second then its frequency becomes 6.

f=6s⁻1

T=1/f=1/6s

ω=2π/T=2.3/1/6s=36radian/s

ANGULAR ACCELERATION

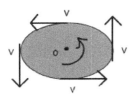

Look at the picture given below.

The speed of the system is constant and we show it with "v". On the contrary direction of the speed changes as time passes and always tangent to the circle. Change in the direction of velocity means system has acceleration, which is called **angular acceleration**. Since the acceleration is;

a= (Vf-Vi)/t

Direction of the acceleration is same as the direction of change of velocities.

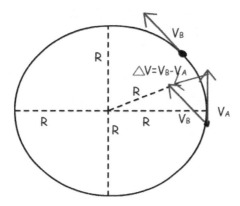

We should find the direction of the changes in the velocity by using vector properties. Let's show how we find the direction of acceleration. Picture shows the change in the direction of velocity. As you can see direction of the resultant velocity vector is towards to the center of the circle. Because of the direction of acceleration, we call it **centripetal acceleration**.

Mathematical representation of centripetal acceleration is;

$$a_{centripetal} = -\frac{4\pi^2 r}{T^2}$$

"-"Sign in front of the formula shows the direction with respect to the **R** position vector.

We can rewrite centripetal acceleration in terms of angular velocity and tangential velocity.

$$a_{centripetal} = -\omega^2 r \quad \text{or,} \quad a_{centripetal} = \frac{v^2}{r}$$

Example: If the tangential speed of the object is 3m/s, which is doing circular motion on a path of radius 2m, find the centripetal acceleration of it.

$$a_{centripetal} = -\frac{4\pi^2 r}{T^2}$$

Or we can rewrite it as;

a=V²/r=(3m/s)²/2m=4,5m/s²

CENTRIPETAL FORCE

So far we have talked about angular speed, tangential speed and centripetal acceleration. As I mentioned in Newton's Second Law of motion, if there is a net force than our mass has acceleration. In this case we find the acceleration first, so if there is acceleration then we can say there must be also a net force causing that acceleration. The direction of this net force is same as the direction of acceleration that is towards to the center. Look at the given picture that shows the directions of force and acceleration of an object doing circular motion in vertical.

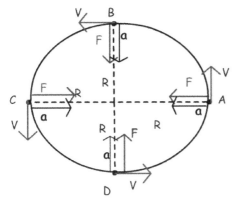

Don't forget! Direction of acceleration and force is always same.

From the Newton's Second Law of Motion;

F=m.a
Fc=-m4π²r/T² or   Where; m is mass of the object, r is the radius of the circle, T is period, V is the tangential speed

Fc=mv²/r

Look at the given examples of centripetal force.

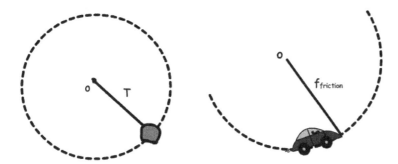

First picture shows the motion of a stone tied up with a string doing circular motion. T represents the tension of the string towards to the center. In this case centripetal force is equal to the tension in the rope. In second picture, a car has circular motion.

Force exerted by the friction to the tiers of the car makes it do circular motion. Only force towards to the center is friction force. Thus, in this case our centripetal force becomes the friction force. We can increase the number of examples. For example, electrical forces or gravitational force towards to the center can be centripetal force of that system.

Example: Two objects A and B do circular motion with constant tangential speeds. Object A has mass 2m and radius R and object B has mass 3m and radius 2R. If the centripetal forces of these objects are the same find the ratio of the tangential speed of these objects.

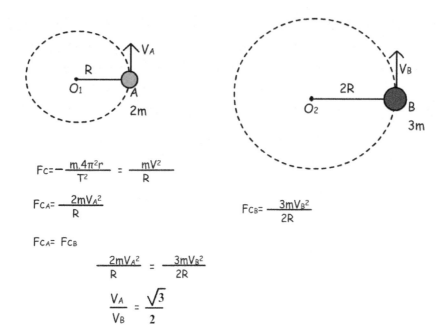

$$F_c = \frac{m \cdot 4\pi^2 r}{T^2} = \frac{mV^2}{R}$$

$$F_{CA} = \frac{2mV_A^2}{R} \qquad F_{CB} = \frac{3mV_B^2}{2R}$$

$$F_{CA} = F_{CB}$$

$$\frac{2mV_A^2}{R} = \frac{3mV_B^2}{2R}$$

$$\frac{V_A}{V_B} = \frac{\sqrt{3}}{2}$$

Example: A car makes a turn on a curve of having radius 8m. If the car does not slide find the tangential velocity of it. (Coefficient of friction between the road and the tiers of the car =0, 2 and g=10m/s²)

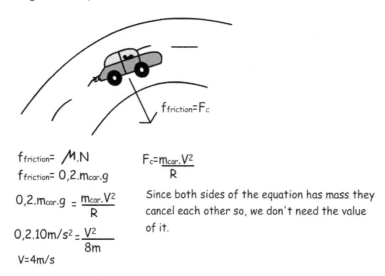

$$f_{friction} = \mu \cdot N \qquad F_c = \frac{m_{car} \cdot V^2}{R}$$

$$f_{friction} = 0{,}2 \cdot m_{car} \cdot g$$

$$0{,}2 \cdot m_{car} \cdot g = \frac{m_{car} \cdot V^2}{R}$$

Since both sides of the equation has mass they cancel each other so, we don't need the value of it.

$$0{,}2 \cdot 10 m/s^2 = \frac{V^2}{8m}$$

$$V = 4 m/s$$

CIRCULAR MOTION ON INCLINED PLANES

We examine this subject with an example. Look at the given picture and analyze the forces shown on the picture.

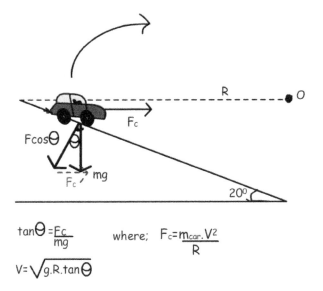

$$\tan\theta = \frac{F_c}{mg} \quad \text{where;} \quad F_c = \frac{m_{car} \cdot V^2}{R}$$

$$V = \sqrt{g \cdot R \cdot \tan\theta}$$

As you can see from the picture given above, we showed the forces acting on the car. For having safe turn on the curve car must have the value given above which is the top limit. It can also have less speed than given above. If we want to increase the speed of the turn we should increase the slope of the road.

Example: Car having mass 1500kg makes a turn on the road having radius 150m and slope 20°. What is the maximum speed that car can have while turning for safe trip?

maximum velocity that car can have for safe trip = $\sqrt{g \cdot R \cdot \tan\theta}$

$$V = \sqrt{10 m/s^2 \cdot 100m \cdot \tan 20°}$$

$$V = 6\sqrt{10} \, m/s$$

CENTRIFUGAL FORCE

In daily life we feel a force on us while we are in a system doing circular motion. For instance, when a car goes around a curve we feel that as if something pulls us outward to the center of that curve. In real, of course there is no such a force exerting on us. In previous topic we explained the centripetal force and gave some examples to it.

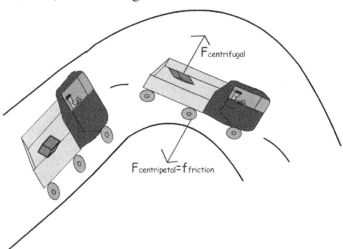

Now, we use again same examples to clarify centrifugal force, which means "outward from the center". Look at the given picture and explanations of centrifugal force;

A truck moving in a straight line carries a box. During linear motion box does not move and has the same velocity with the truck. However, when the truck goes around a curve, box starts to move outward to the center of the curve as if an unknown force pulls it. We also feel this force when in a bus or car while it is doing circular motion. Is there a force pulling us outward from the center? The answer is of course NO! Let me explain this complex situation with Newton's Laws of Motion. We have said that for having acceleration there must be an unbalanced net force on that system. Here, a friction force between the road and the tiers of the truck becomes our unbalanced net force. It changes the direction of motion and truck does circular motion. On the other hand, the friction between the box and the surface of the truck is not enough to make it does also circular motion. Because of the Newton's First Law of Motion "Law of Inertia", box tends to move in a straight line. Thus, it slides on the truck and feels like something pulls it outward from the center. This is only inertia of the box; in real there is no centrifugal force.

TORQUE

We define torque as the capability of rotating objects around a fixed axis. In other words, it is the multiplication of force and the shortest distance between application point of force and the fixed axis. From the definition, you can also infer that, torque is a vector quantity both having direction and magnitude. However, since it is rotating around a fixed axis its direction can be clockwise or counter clockwise. During the explanations and examples we give the direction "+" if it rotates clockwise direction and "-" if it rotates counter clockwise direction. Torque is shown in physics with the symbol "τ". You can come across torque with other name "moment".

Now, we examine given pictures one by one to understand torque in detail. How can we find the shortest distance between the applied force and fixed axis? All you know that, shortest distance between two points is the straight line connecting them.

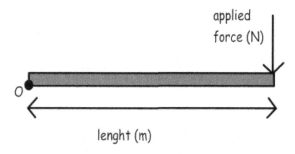

In this situation distance connecting these two points is the length of the object. Direction of the torque is "+" because force rotates the object in clockwise direction. (We ignore the weight of the object in all situations given above.)

Thus, we can write the torque equation like;

τ=Applied Force.Distance

In this picture, we have a different situation where the object is fixed to the wall with an angle to the horizontal.

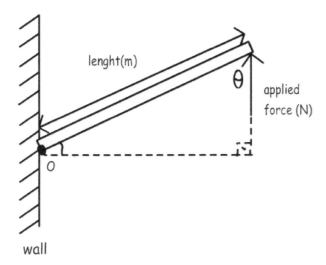

Direction of the torque in this situation is "-" because force rotates the object in counter clockwise direction. As we said before we need shortest distance between the force and turning point. Dashed line in the picture shows this distance that can be found by using trigonometry. Final equation of torque becomes;

τ =Applied Force.Distance.sinθ

Final situation shows that, if the extension of the force passes on the rotation axis then what will be the torque?

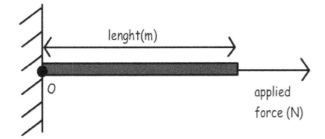

I want to explain this situation by giving other example. Think that you are opening a door. If you push the door as in the case of given picture above, the door does not move. However, if you apply a force to the door like in the first and second situations given above the door is opened or closed. What I try to say is, if the force is applied to the turning point then it does not rotate the object and there won't be torque.

Example: If the given triangle plate is fixed from the point O and can rotate around this point, find the total torque applied by the given forces.

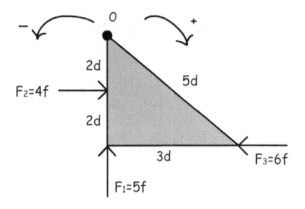

We find the torques of the forces one by one and finally sum them up considering their directions.

$T_1 = F_1 \cdot d_1 \cdot \sin\theta \qquad T_1 = 5f \cdot 0 = 0$

$T_2 = F_2 \cdot d_2 \cdot \sin\theta \qquad T_2 = (-)4f \cdot 2d = -8fd$

$T_3 = F_3 \cdot d_3 \cdot \sin\theta \qquad T_3 = 6f \cdot 4d = 24fd$

Total torque;

$T_{total} = T_1 + T_2 + T_3 = 0 + (-)8fd + 24fd = 16fd$ to the clockwise direction

Example: If the plate is fixed from the point O, find the net torque of the given forces.

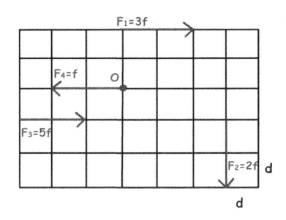

$T_1 = F_1 \cdot d_1 \cdot \sin\theta \qquad T_1 = 3f \cdot 2d = 6fd$

$T_2 = F_2 \cdot d_2 \cdot \sin\theta \qquad T_2 = 2f \cdot 3d = 6fd$

$T_3 = F_3 \cdot d_3 \cdot \sin\theta \qquad T_3 = (-)5f \cdot d = -5fd$

$T_4 = F_4 \cdot d_4 \cdot \sin\theta \qquad T_4 = f \cdot 0 = 0$

Total torque;

$T_{total} = T_1 + T_2 + T_3 + T_4 = 6fd + 6fd + (-)5fd + 0 = 7fd$ to the clockwise direction

MORE EXAMPLES RELATED TO ROTATIONAL MOTION

Example: An object, attached to a 0,5m string, does 4 rotations in one second.

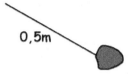

Find period, tangential velocity and angular velocity of the object.

If the object does 4 rotations in one second, its frequency becomes;

f=4s^{-1}

T=1/f=1/4s

Tangential velocity of the object;

V=2.π.f.r

V=2.3.4.0,5

V=12m/s

Angular velocity of the object

ω=2.π.f=2.3.4=24radian/s

Example: Find the relation between tangential and angular velocities of points X, Y and Z.

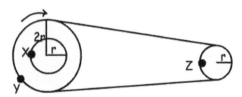

X and Y rotate together, so if X does one rotation then Y also does one rotation. On the contrary, if Y does one revolution, Z does two revolutions.

Angular velocities of the X, Y and Z are;

ω$_X$=ω$_Y$=ω$_Z$/2

Example: An object hanged on a rope L=0,5m, does rotational motion. If the angle between rope and vertical is 37^0, find the tangential velocity of the object. (g=10m/s², cos37^0=0,8, sin37^0=0,6)

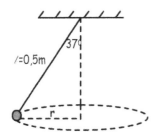

Free body diagram of system is given below;

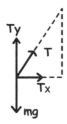

Horizontal component of tension on the rope makes object rotate.

T$_X$=mV²/r, T$_Y$=m.g

Radius of the motion path is;

r=L.sin37^0=0,5.0,6=0,3m

tan37^0=T$_X$/T$_Y$

3/4=mV²/r/m.g

3/4=V²/g.r

V=3/2m/s

Example: A device does 1800 revolution in 6 minutes. Find angular velocity and period of this device. (π=3)

1800.T=6.60s

T=360/1800=1/5s

ω=2π/T=2.3/1/5=30 radian/s

Example: Friction constant of a circular road, having radius 50m, is 0,2. Find the initial velocity of the car having 1500 kg to turn junction safely.

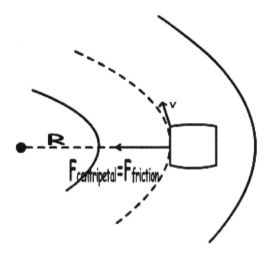

$F_{friction} \geq F_{centripetal}$ **(for safe turn)**

k.m.g≥m.V²/r

k.g.r≥V²

100≥V²

10m/s≥V or

36km/h≥V

Example: A marble is thrown with velocity V from point A. If no force is exerted on surface by marble at point B, find the force exerted on point A by marble. (Assume that velocity of marble is constant.)

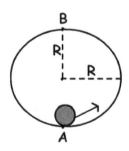

Since no force is exerted on point B by marble;

mg=mV²/R

V²=g.R

Force exerted by marble on point A;

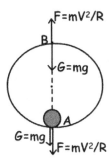

$F_A = G + F$

$F_A = mg + mV^2/R$ (we put $V^2 = gR$ into the equation)

$F_A = mg + mgR/R$

$F_A = 2mg$

Example: Two objects rotate with same frequency around point O. One of the objects has mass m and other one has mass 9m. If centripetal forces exerted on objects having mass m is F_1, and object having mass 9m is F_2, find ratio of F_1/F_2.

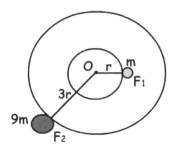

$V_1 = 2.\pi.r.f$

$V_2 = 2.\pi.3r.f$

$V_2 = 3V_1$

$F_1/F_2 = mV_1^2/r / 9mV_2^2/3r = V_1^2/r . 3r/9V_2^2$

$F_1/F_2 = 1/27$

OPTICS

Optic is one of the branch of physics, which deals with the light, and properties of it. We know that light shows both the particle and wave characteristics. However, in this unit we will learn the particle characteristics of the light. Some of the topics will be covered in this unit are; reflection, refraction of light, plane mirrors, concave and convex mirrors, reflection of light from the mirrors, prisms and behavior of light in different mediums.

PROPERTIES of LIGHT

Light shows both wave and particle characteristics.

Light can travel in vacuum and speed of it in vacuum is 300.000.000m/s

Light consists of particles, which are called as photons.

People see objects with the help of light.

Light travels in straight light and during travel it chooses the way takes least time. This principal is called Fermat's Principle.

Light interacts with the matter and according to the types of material it reflects or refracts.

REFLECTION of LIGHT

Reflection is the turning back of the light from the surface it hits. Incoming and reflected lights have same angle with the surface. If the surface reflects most of the light then we call such surfaces as **mirrors**. Examine the given pictures below. They show regular and diffuse reflection of light from given surfaces.

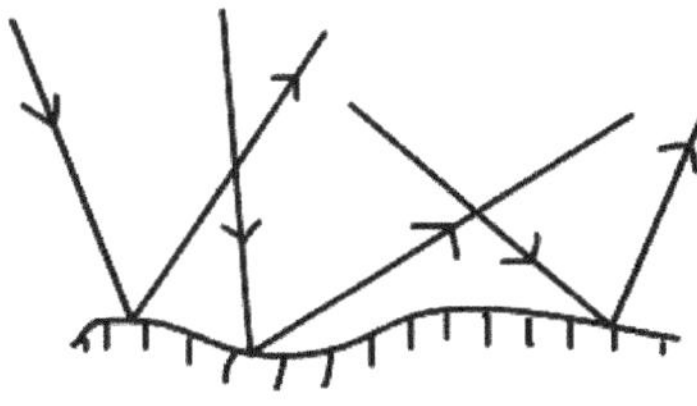

Diffuse reflection from rough surfaces

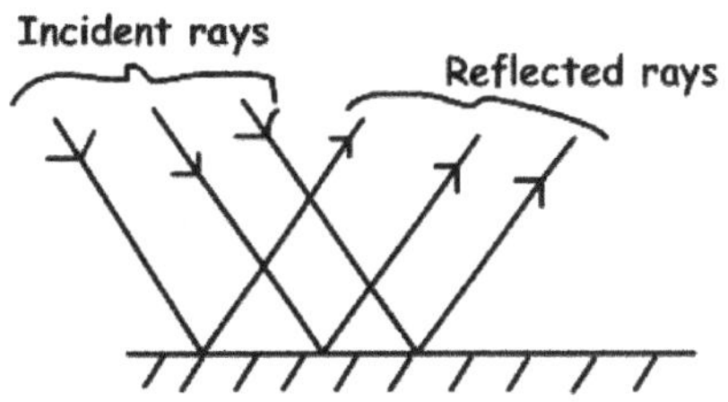

Regular reflection from smooth surfaces

Laws of Reflection

First law of reflection states that; Incident ray, reflected ray and Normal to the surface lie in the same plane.

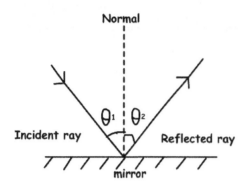

$\theta_1 = \theta_2$
angle of incident ray = angle of reflection

Angle of incident ray is equal to the angle of reflection ray.

Example: Find the angle of incident ray from the given information in picture?

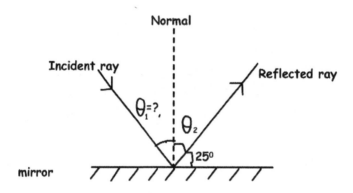

We know that angle of incident ray is equal to the angle of reflected ray and normal of the system is perpendicular to the surface. Thus,

$\theta_2 + 25^0 = 90^0$
$\theta_2 = 65^0$ $\qquad \theta_1 = 65^0$

PLANE MIRRORS AND IMAGE FORMATION IN PLANE MIRRORS

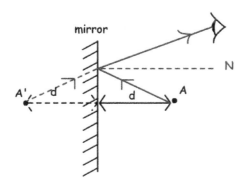

If the reflecting surface of the mirror is flat then we call this type of mirror as **plane mirrors.** Light always has regular reflection on plane mirrors. Given picture below shows how we can find the image of a point in plane mirrors.

We have to see rays coming from the object to see it. If the light first hits the mirror and then reflects with the same angle, the extensions of the reflected rays are focused at one point behind the mirror. We see the coming rays as if they are coming from the behind of the mirror. At point A' image of the point is formed and we call this image **virtual image** which means not real. The distance of the image to the mirror is equal to the distance of the object to the mirror. If we want to draw the image of an object in plane mirrors we follow the given steps below. First look at picture and then follow the steps one by one.

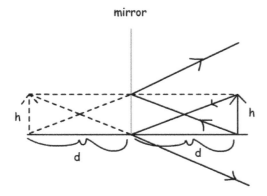

In plane mirrors, we use the laws of reflection while drawing the image of the objects. As you see from the picture we send rays from the top and bottom of the object to the mirror and reflect them with the same angle it hits the mirror. The extensions of the reflected rays give us the image of our object. The orientation and height of the image is same as the object. In plane mirrors always virtual image is formed.

Example: Find the image of the given object.

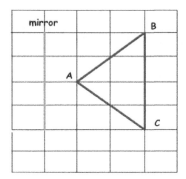

As I said before, image of the object is formed behind the mirror with the same distance of object. We draw first point A' which is the image of point A, we placed it one unit away from the mirror, then points B' and C' are placed with the same way. We connect these 3 points and the image of object becomes formed. The dashed line in the left side of the mirror is our image.

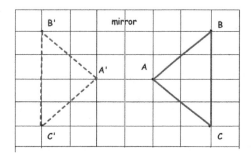

Moving Plane Mirror

If the plane mirror is moving with a velocity v, what will be the velocity of the image? Is it also moving with the same velocity of the mirror and what is the direction of the images velocity? We will try to answer these questions in this section.

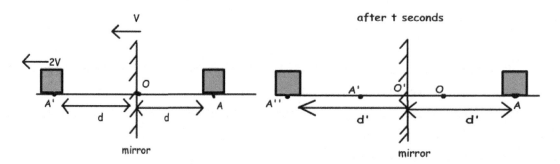

If the mirror moves with a velocity of **V** then image of the object moves with the velocity of **2V** with the same direction of mirror. Second picture shows the locations of the mirror and image after **t** seconds. As it seen from the picture, distance traveled by the image is twice of the distance traveled by the mirror since the velocity of the image is **2V**.

Now, we will examine a situation in which mirror is a stationary and object moves with a velocity of **V**.

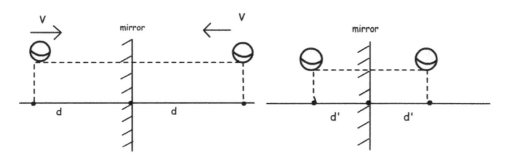

If the object moves with a velocity of **V** then image also moves with same speed but opposite in direction. Pictures given above show the motion of object and image. Second picture is the representation of locations after **t** seconds.

Example: Find the velocity of the image of the car with respect to ground if both the mirror and car moves with a given velocities.

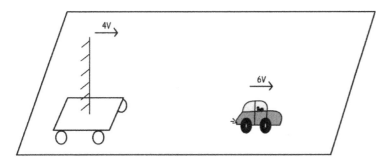

If car moves with velocity 6V then image moves with -6V, we put "–"sign since image moves to the left. If mirror moves with velocity 4V to the right, then image of the car moves with the velocity 8V to the right.

Velocity of the image=8V+ (-6V) =2V to the right with respect to ground

CURVED MIRRORS

We call these types of mirrors also spherical mirrors because they are pieces of a sphere. If the reflecting surface of the mirror is outside of the sphere then we call it **convex mirror** and if the reflecting surface of it is inside the sphere then we call it **concave mirror**. There are some fundamental terms we should learn before we pass to the ray diagrams and image formation in curved mirrors such as principal axis, focal point, center of curvature, radius of curvature and vertex.

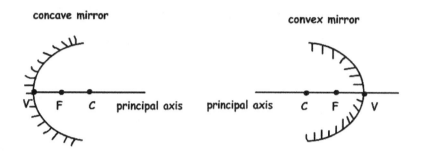

Center of Curvature:

As you can understand from the name it is the center of the sphere, which the mirror is taken from. It is denoted with "**C**" in the diagrams.

Principal Axis:

Line coming from the center of the sphere to the mirror is called as principal axis.

Vertex:

It is the intersection point of the mirror and principal axis. We show it with letter "**V**" in ray diagrams.

Focal Point:

For concave mirrors and thin lenses rays coming parallel to the principal axis reflects from the optical device and pass from this point. For convex mirrors and thick lenses rays coming from this point or appear to coming from this point reflect parallel to the principal axis from the optical device. Another explanation for this term is that, it is the point where the image of the object at infinity is formed. It is denoted with the letter **F** or sometimes **f** in ray diagrams.

Radius of Curvature:

It is the distance between center of the sphere and vertex. We show it with **R** in ray diagrams.

Concave Mirrors

We give the definition of concave mirrors in previous sections. Now we will examine the reflection of light from this type of mirrors and image formation in concave mirrors. Let's start with the reflection of light with special examples.

In this example, ray coming parallel to the principal axis reflects from the mirror and passes from the focal point. As in the case of plane mirrors light obey the law of reflection. In this case normal of the mirror is the line coming from the center of the mirror. If you want to draw the reflection of any light you can use this technique. Draw your normal and with the same angle of incident ray reflect your light.

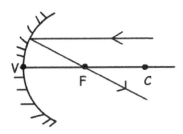

Ray coming from the focal point reflects from the mirror and goes parallel to the principal axis, as in the first situation.

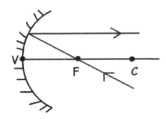

The ray coming from center turns back on itself after reflecting from the mirror. As I said before line coming from the center of the mirror is the normal of that system and hits the mirror perpendicularly.

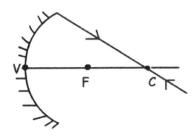

Ray hitting the mirror at vertex point reflects with the same angle it comes.

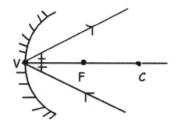

If the ray is not one of the special rays given above, then you should draw the normal of the system and reflect the light with the same angle as it hits the mirror.

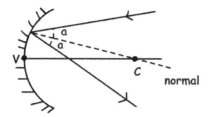

Example: Find the relation between the focal lengths of the mirrors given below.

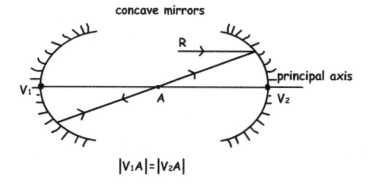

$|V_1A|=|V_2A|$

Light ray R comes parallel to the principal axis and reflects from the mirror 2 and passes from the point A. We know that rays coming parallel to the principal axis pass from the focal point after reflection. Thus, point A is the focal point of the mirror 2.

Ray coming from point A turns back on itself after reflecting from mirror 1, we know that rays coming from the center of curvature turns back on itself. So, point A is the center of curvature of the mirror 1.

$|V_1A| = 2f_1$ and $|V_2A|=f_2$

$2f_1=f_2$

Example: Look at the given system, there is a concave mirror and plane mirror located on the focal point of the concave mirror. Draw the path of the ray R and show how it leaves the system.

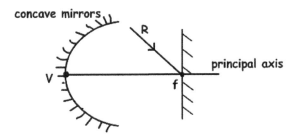

Ray coming to the plane mirror reflects with the same angle, reflected ray from the plane mirror comes to the concave mirror from its focal point and after reflecting from the concave mirror it travels parallel to the principal axis. It comes to the plane mirror with an angle 90°, it turns back on itself and follows the same path and leaves the system.

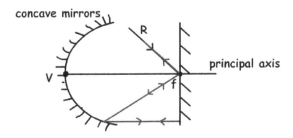

Image Formation in Concave Mirrors

While drawing images of the object we benefits from the special rays given above. We use them because we know the paths of them. Let's start drawing images of the objects located in different parts of the mirror.

1. If the object placed at the center of the mirror, image is also formed at center, real, inverted and with the same size as object. To find this image we send two rays from the top of the object. One of them is parallel to the principal axis, which passes from the focal point after reflection and second ray passes from the focal point and goes parallel to the principal axis after reflection. The intersection point of these two reflected rays gives us the location of image. As you can see from the picture green one is image of the object.

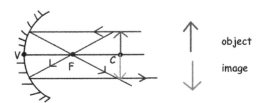

2. If the object placed away from the center of the mirror, image is formed between the focal point and center of the mirror. Properties of image are, real, inverted and reduced in size.

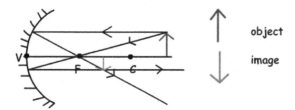

3. If the object placed between the focal point and center of the mirror, then image is formed away from the center. Characteristics of the image are; real, inverted and magnified in size.

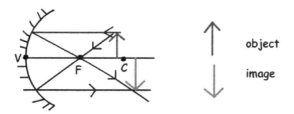

4. In this picture object is placed at focal point and as you can see reflected rays goes parallel to each other. In other words, they do not intersect in any point, thus we assume that image is formed at infinity.

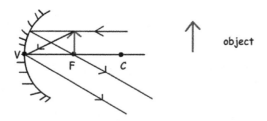

5. If the object placed between the focal point and vertex then **virtual** image is formed behind the mirror. Since the rays reflected from the mirror do not intersect, their extensions behind the mirror intersect and virtual image is formed. Image is magnified in size and erect unlike the real images.

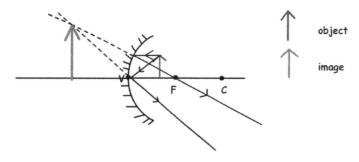

Example: We have an optical system including a concave mirror and a plane mirror in the picture given below. Object is located at point A; its first image is formed in plane mirror and second one is in concave mirror. Find these images of the object.

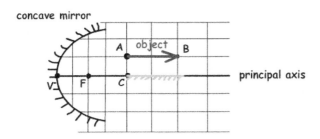

Image of the object from the plane mirror is located at A' B' points

The image located at A'B' is object for concave mirror. We use mirror equations to find final location and height of the image.

$$\frac{1}{f} = \frac{1}{D_o} + \frac{1}{D_i}$$

We use this formula for points A' and B'

Since A' is located at the center of the mirror, its image also located at the center inverted and real.

$$\frac{1}{2} = \frac{1}{6} + \frac{1}{D_i}$$

$D_i = 3$

$$\frac{D_o}{D_i} = \frac{H_o}{H_i} = 2$$

So, height of the image is half of the real height. It is given in blue in the picture given above.

Example: Look at the given picture, if the distance between the object and its image is 120cm; find the focal length of this mirror. Image of the object is behind the mirror, erect and its length is three times larger than the object.

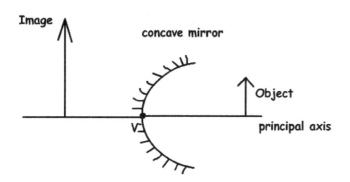

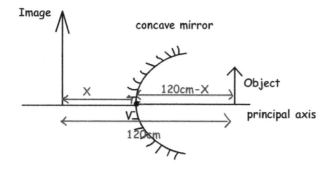

where;
Hi=height of the image
Ho=height of the object
Di=distance between image and mirror
Do=distance between object and mirror

Hi=3Ho

$$\frac{Hi}{Ho} = \frac{Di}{Do} = 3 \quad \text{Where, Di=X and Do=120-X}$$

Di=3Do
X=3(120-X)
X=Di=90cm
Do=30cm

$$\frac{1}{f} = \frac{1}{Do} + \frac{1}{Di}$$

$$\frac{1}{f} = \frac{1}{30cm} - \frac{1}{90cm} \quad \text{(Since image is formed behind the mirror we put "-" sign in front of it.)}$$

f=45cm

Convex Mirrors

We give the definition of convex mirrors in previous sections. Now we will examine the reflection of light and image formation in convex mirrors. Let's start with the reflection of light with special examples.

In convex mirrors, ray coming parallel to the principal axis goes after reflection as if it comes from the focal point of the mirror.

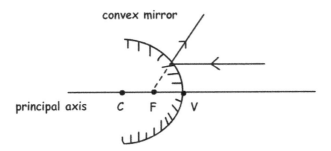

If the extension of the coming ray passes from the focal point, it travels parallel to the principal axis after reflection from the mirror.

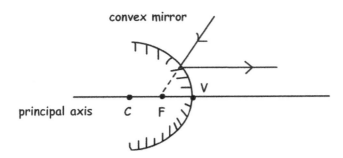

If the extension of coming ray passes from the center of the curvature, ray reflects from the mirror and travels on itself. As we said before line joining the mirror to the center of the curvature is perpendicular to the mirror and normal of the system.

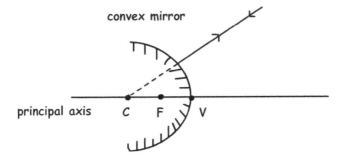

Ray strikes to the mirror at vertex, reflects from the mirror with the same angle it comes because, principal axis is the normal of the system.

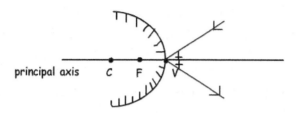

Example: Look at the path of the ray R in the given picture, **f₁** is the focal length of the convex mirror and **f₂** is the focal length of the concave mirror. Find the distance between mirrors in terms of **f₁** and **f₂**.

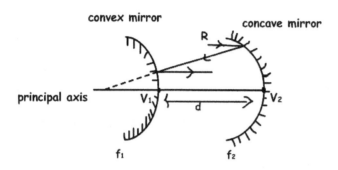

Ray coming parallel to the principal axis must go to the focal point of the concave mirror, after reflecting from the concave mirror ray strikes to the convex mirror and goes parallel to the principal axis. This shows us that; ray coming from concave mirror goes to the focal point of the convex mirror. Let me show these explanations on the picture.

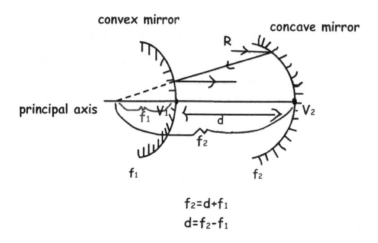

$$f_2 = d + f_1$$
$$d = f_2 - f_1$$

Image Formation in Convex Mirrors

While drawing images of the object we benefits from the special rays given above. We use them because we know the paths of them. In convex mirrors image is usually formed behind the mirror, it is virtual and erect. Location of the image is always between the focal point and vertex of the mirror. Look at the given pictures below, they show what I try to say.

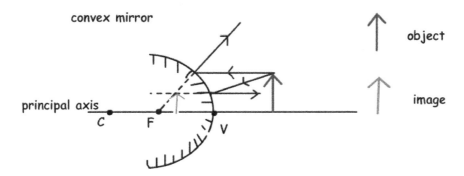

We send two rays from the top of the object. One of the rays is parallel to the principal axis and reflects from the mirror as it comes from focal point and the extension of the second ray passes from the focal point so it travels parallel to the principal axis after reflection. Behind the mirror, extensions of the two reflected rays intersect at one point at which image of the object is located. As you can see from the picture height of the image is smaller than the object, it is virtual and erect. Now look at another example given below.

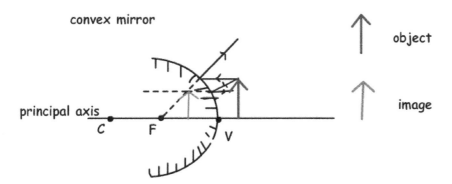

In this picture, object is closer to the mirror. Image is again formed between the focal point and vertex. However, in this case height of the image is larger than the situation given above. We can conclude that, closer the object to the mirror bigger the height of the image. But, don't forget, image is always smaller than the object.

The last example is given below in which object is at infinity and image is formed behind the mirror at focal point like a point.

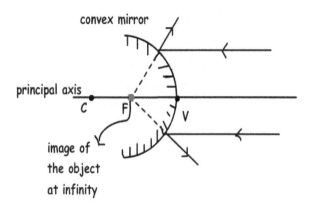

Example: Plane mirror and convex mirror are placed on same principal axis. If the focal length of the convex mirror is 12 cm find the distance d shown in the picture.

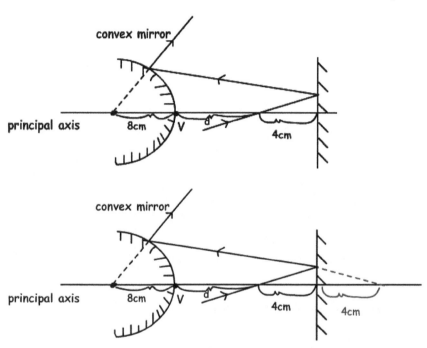

we use the mirror equations to solve this problem.

$$-\frac{1}{f} = \frac{1}{D_o} - \frac{1}{D_i}$$

we put "-" sign in front of the focal length and image distance since they are behind the mirror we take them as negative.

$$-\frac{1}{12cm} = \frac{1}{d+8cm} - \frac{1}{8}$$

red dashed line shows where the light comes from, so we take Di=d+8

d=16cm

Mirror Equations of Curved Mirrors

We use the given picture below to derive the equations of concave and convex mirrors.

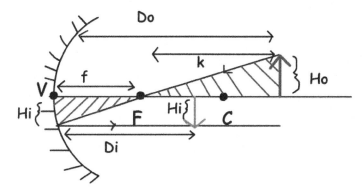

Where;
Ho: Height of the object
Hi: Height of the image
Do: Distance between the object and mirror
Di: Distance between the image and mirror
f: Focal length
k: Distance between object and focal point

While deriving equations we use the similarities of triangles given picture above. We show them with red lines in the picture. I do not want to make confusion in your mind and write down the equations that I get from similarity of two triangles.

1. $\dfrac{Ho}{Hi} = \dfrac{f}{k}$

2. $\dfrac{Ho}{Hi} = \dfrac{Do}{Di}$

we get final equation by using 1 and 2

$$\dfrac{1}{f} = \dfrac{1}{Do} + \dfrac{1}{Di}$$

In this equation, we take the quantities positive if they in front of the mirror. In other words, if the object, image and focal point of the mirror are located in front of the mirror (like concave mirrors) we take them positive; if they are located behind the mirror (like convex mirrors) we take them negative. In convex mirrors focal length is taken as negative since it is behind the mirror and in concave mirrors if the image is behind the mirror then we take it also negative.

REFRACTION

As I said before light travels in a straight line in any homogeneous medium. However, when it passes from one medium to another medium it changes its direction. We call this change in the direction of light **refraction**.

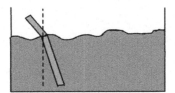

Since the densities of the mediums are different light travels with different speed in different mediums. Speed of light in vacuum is 300.000.000km per hour. You see the stick as it is broken in the given picture. It is a clear example of refraction of light. We find the amount of refraction by using the refractive indexes of the mediums. What is the refractive index of a medium? It is the ratio of the speed of light in vacuum to the speed of the light in given medium. Refractive index of medium A is given below;

$$n_a = \frac{\text{Speed of the light in vacuum}}{\text{Speed of light in a medium}} = \frac{c}{V_a}$$

Now we examine the refraction of light from two mediums. We assume the refractive index of air 1.

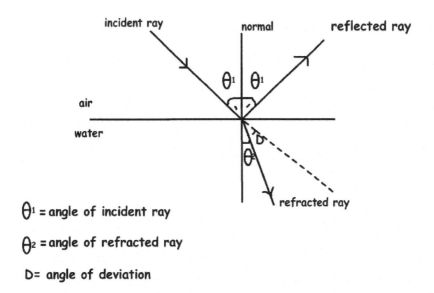

θ_1 = angle of incident ray

θ_2 = angle of refracted ray

D = angle of deviation

The drawing given above shows the angles of reflected and refracted rays. You may some confusion in your mind about the reflection of the light in refracting. However, I want to mention it here, when an incident ray passes one medium to another some part of it is reflected from the surface of boundary and the rest is refracted. The amount of reflection is related to the difference between the refractive indexes of the mediums. Later in this topic we will see it in detail.

The Laws of Refraction

Incident ray, reflected ray, refracted ray and the normal of the system lie in the same plane.

Incident ray, coming from one medium to the boundary of another medium, is refracted with a rule derived from a physicist Willebrord Snellius. He found that there is a constant relation between the angle of incident ray and angle of refracted ray. This constant is the refractive index of second medium relative to the first medium. He gives the final form of this equation like;

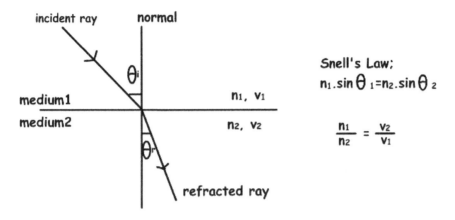

Snell's Law;
$n_1 \cdot \sin\theta_1 = n_2 \cdot \sin\theta_2$

$$\frac{n_1}{n_2} = \frac{v_2}{v_1}$$

Where n_1 is the refractive index of first medium and n_2 is the refractive index of second medium, v_1 is the speed of light in firs medium and v_2 is the speed of light in second medium.

Example: Find the velocity of the ray in a medium having refractive index 2.

$$n = \frac{c}{v}$$

$$2 = \frac{300.000 \text{ km/h}}{v} \qquad v = 150.000 \text{ km/h}$$

Example: A ray coming from medium X is refracted as shown in the figure below while passing to the medium Y. Find ratio of the refractive indexes of the mediums. (Sin37°=0, 6 and sin53°=0, 8)

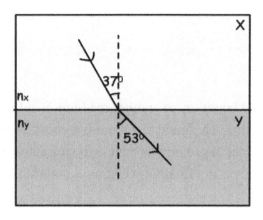

We use snell's law to solve this example
Snell's Law;
$n_1 \cdot \sin\theta_1 = n_2 \cdot \sin\theta_2$
$n_x \cdot \sin 37° = n_y \cdot \sin 53°$
$n_x \cdot 0,6 = n_y \cdot 0,8$

$$\frac{n_x}{n_y} = \frac{4}{3}$$

There are some key points that help you to clarify some concepts in your mind like refraction, refractive index and velocity of light in different mediums.

* When we give refractive index of medium you should understand that it is the relative refractive index of that medium.

* In general, if the density of medium increases than refractive index of that medium also increases. However, of course there are some exceptions like water.

* Velocity of the light in a medium is inversely proportional to the refractive index of that medium. If the refractive index increases then velocity of the light decreases.

* If the light comes perpendicular to the boundary of two different mediums, it does not change its direction because it is on the normal line of the system. But, the velocity of the light changes since the density of the medium changes.

* Refractive index of the medium is also depends on the color of the coming light. For example, refractive index of the medium for violet colors is larger than the refractive index of the medium for other colors.

The angle of refraction of the light coming from the medium having smaller refractive index is smaller than the angle of incident ray. Look at the given diagram that shows this relation.

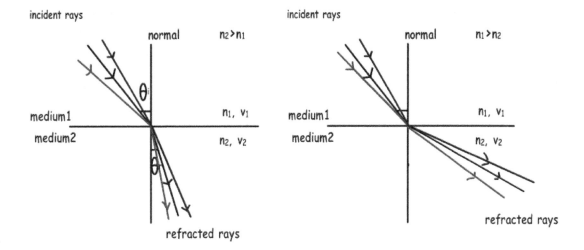

Example: Find the path of the ray after refracting.

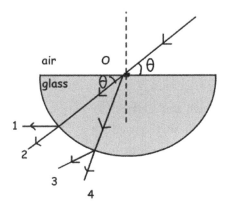

Ray coming from air to glass is refracted from the surface, it cannot travel with same angle in glass, and thus 1 and 2 choices are eliminated. When the light comes to the boundary of the glass it does not refract since it comes from the normal of the system and goes in a straight line. Thus, it follows the path shown in 4.

CRITICAL ANGLE AND TOTAL REFLECTION

Look at the given picture. Rays with different angles coming from the source at the bottom of the water filled glass, refracted from the surface, reflected and directly pass to the second medium without any reflection and refraction.

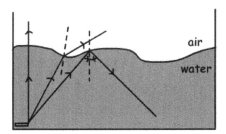

In the previous section we have learned Snell's law of refraction. According to this law; reflecting, refracting or directly passing of light to the other medium is depends on angle of incident ray and refractive indexes of the mediums. Here, you see behavior of light coming from the medium having larger refractive index and try to pass to the medium having smaller refractive index. As you can see as we change the angle of incident ray behavior of light changes. First light directly passes to the second medium without any reflection or refraction, second light refracts and final light reflects from surface. According to the angle of incident ray, at one point it does not refract but goes parallel to the boundary of the mediums. We called this angle as **critical angle**. If the angle of incident ray is larger than the critical angle then it does not refract but it does total reflection. In the picture given above, first light is coming from the normal thus it does not reflect or refract; it passes directly to the second medium. Second light coming with an angle smaller than the critical angle, refracts and passes to the second medium. However, third light coming with an angle larger than the critical angle does total reflection and cannot pass to the second medium. Now we use Snell's law of refraction and find the equation of critical angle.

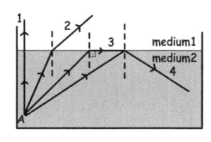

Snell's Law;
$n_1 \cdot \sin\theta_1 = n_2 \cdot \sin\theta_2$

Third light comes with and critical angle and goes parallel to the surface. We take the angle of refration for this ray $90°$.

Snell's Law;
$n_2 \cdot \sin\theta_c = n_1 \cdot \sin 90°$ where θ_c is the critical angle

$\sin\theta_c = \dfrac{n_1}{n_2}$

θ_c

Where θ_c is critical angle and rays coming with an angle larger than θ_c make total reflection.

Example: Which paths, given in the picture below, the light ray A can follow?

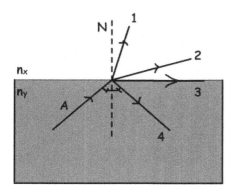

We do not know the refractive indexes of the mediums. If **nx** is larger than **ny** then light ray follows the path shown in 1. It approaches to the normal. On the contrary, if the **ny** is larger than **nx** and angle of incident ray is smaller than the critical angle then it follows the path shown in 2. If the angle of incident ray is equal to the critical angle and **ny** is larger then **nx** the ray A follows the path shown in 3. Finally, if the angle of incident ray is larger than the critical angle and **ny** is larger than **nx** ray follows the path shown in 4 it does total reflection.

Example: Refractive index of medium A is 2, and medium B is 1,6. Find the critical angle of the rays coming from the medium A to B.

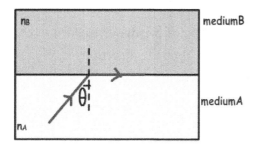

Snell's Law;

$n_B \cdot \sin \theta_c = n_A \cdot \sin 90°$ where $\sin 90° = 1$

$\sin \theta_c = \dfrac{n_B}{n_A} = \dfrac{1,2}{2} = 0,6$ which corresponds to the angle $37°$

TOTAL REFLECTION IN PRISMS

Critical angle for glass in air is 42°. Thus, rays coming with an angle 45°, which is larger than the critical angle, to the glass make total reflection.

This prism is used for to change the direction of incident ray 90°. Rays come to the prism perpendicularly and they do not refract, they hit the surface with an angle 45° and make total reflection.

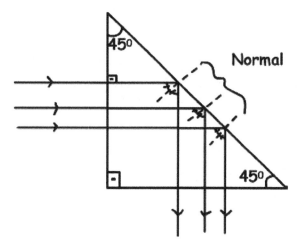

In this prism, rays again make total reflection. This type of prisms is used for to change direction of rays 180°. As you can see rays leave the prism in opposite direction.

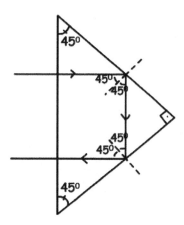

This is the last example of prism that I choose for you. In this case direction of light is not changed it goes parallel to the incident ray after making total reflection.

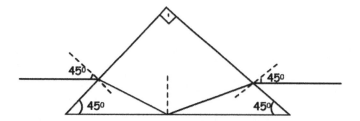

APPARENT DEPTH REAL DEPTH

Picture shows the difference between real depth and apparent depth of the object under water.

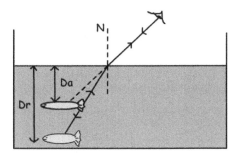

We see objects closer than their real depth to the surface. We can see them only if rays coming from them reaches our eyes. In this picture, ray coming from the fish reaches the observer's eye after refraction. Thus, observer sees the image of the fish at the distance Da from the surface, which is the apparent depth of the fish. On the contrary fish sees the objects away from their real distances. These are all results of the refraction of light. Following diagram helps us to calculate this apparent depth of the object under different mediums.

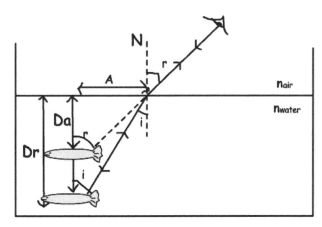

Da=Apparent depth
Dr=Real depth
i=angle of incident ray
r=angle of refracted ray

Snell's Law;
$n_{water} \cdot \sin\theta_i = n_{air} \cdot \sin\theta_r$

$n_{water} = \dfrac{\sin\theta_r}{\sin\theta_i}$ for small angels $\sin\theta_i = \tan\theta_i$ and $\sin\theta_r = \tan\theta_r$

writing down the tangents of the angle n_{water} becomes;

$n_{water} = \dfrac{\tan\theta_r}{\tan\theta_i}$ where $\tan\theta_i = \dfrac{A}{Dr}$ and $\tan\theta_r = \dfrac{A}{Da}$

$n_{water} = \dfrac{Dr}{Da}$

As you can see refractive index of the water is equal to the ratio of real depth of fish to the apparent depth of it. More general form of our equation is given below;

$$D_a = \frac{D_r \cdot n_{observer}}{n_{object}}$$

where; D_a is the apparent depth,
D_r is the real depth
$n_{observer}$ is the refractive index of the medium of observer
n_{object} is the refractive index of the medium of object.

Example: An observer looks at the water tank and according to him half of the tank is filled with water. If the height of the tank is 180cm, find the real height of the water in the tank. (n_{water}=4/3)

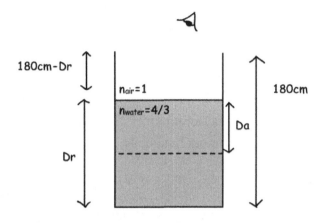

$D_a = D_r \cdot n_{observer} / n_{object}$

$D_a = 3D_r/4$

Observer sees the tank half filled thus;

$3D_r/4 = 180 - D_r$

$D_r = 720/7 = 102,8$ cm and $D_a = 3D_r/4 = 77,1$ cm

MORE EXAMPLES RELATED TO OPTICS

Example: In the picture given below, you see object placed at point A and it's motion at point A'. If we rotate plane mirror 30^0 in clockwise direction, find the final location of image of the object.

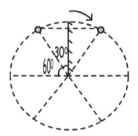

If plane mirror rotates 30^0, then image of the object rotates 60^0. (We have learned this rule in content part. If plane mirror rotates α, then image of the object rotates 2α). Thus, image of the object is;

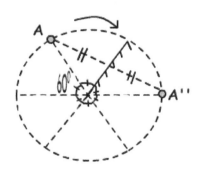

Example: Rays coming from the object A, first reflects from mirror 1 and then reflects from mirror 2. Draw the images of this object.

Image of the object is placed symmetrically behind the plane mirror.

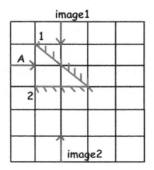

Example: Rays coming from the object AB, first reflects from mirror 1 and then reflects from mirror 2. Draw the images of this object in mirror 2.

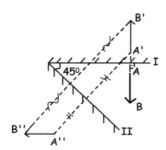

Rays coming from object AB reflects from mirror one and image A'B' is formed behind the mirror I. Rays coming from A'B' reflects from mirror II and image A"B" is formed behind mirror II.

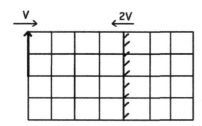

Example: If velocities of the mirror and object with respect to ground are V, and 2V, find the velocity of the image with respect to ground.

Object, mirror and image of the object after t seconds are shown in the figure below. We find the velocity of image using this picture.

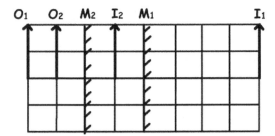

As you can see on the picture given above, image of the object change 5 unit distance, if we say V to 1unit change in t seconds, velocity of the image becomes 5V.

Example: An observer looks at the mirror 1 find the image of the point source with respect to observer.

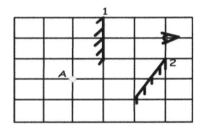

Rays coming from the point source do not hit first mirror directly. They reflect from second mirror and then come to the first mirror. Thus, image of the second mirror becomes object of the first mirror. We draw image of the point source like in the picture given below.

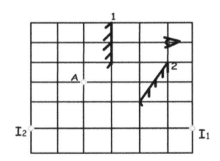

Example: Look at the given picture below. Two concave mirrors are placed on same principal axis. Find focal points of mirror 2 in terms of d.

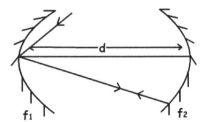

Ray hits the vertex of mirror 1 and reflects with same angle. Ray, coming from first mirror turns back with same path after reflecting from second mirror. Thus, we infer that, first mirror is placed at the center of second mirror. So focal point of the second mirror is;

$d=2f_2$

$f_2=d/2$

Example: Find the relation between the focal points of the concave mirrors given below.

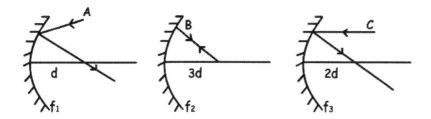

Since extension of the ray A passes from behind of the mirror, f_1 is bigger than d. In the second picture, ray B turns back on itself. Thus; 3d is the center of the mirror and $f_2=3d/2=1,5d$. In third picture, ray coming parallel to principal axis passed from focal point of the mirror. So; $f_3=2d$.

Example: Find shape of image of the object given below.

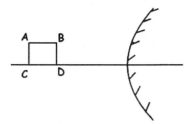

Ray coming from AB parallel to the principle axis reflects from convex mirror and it goes as if it comes from focal point of the mirror. Thus, image of the points A, B are on the extension of reflected ray. AC side of the image is shorter than the BD side since distance between AC and mirror is larger than the distance between BD and mirror. Image of the object is shown in the picture given below;

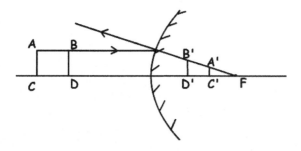

Example: Focal point of the concave mirror given below is 120 cm. An object having length 8 cm is placed at 60 cm away from the mirror. Find the location and height of the image.

$1/f = 1/d_{object} + 1/d_{image}$

$1/120 = 1/60 + 1/d_{image}$

$d_{image} = -120$ cm

"-" Sign in front of image shows that image is imaginary.

$H_{object}/H_{image} = d_{object}/d_{image}$

$8/H_{image} = 60/120$

$H_{image} = 16$ cm

Example: Concave mirror and convex mirror having equal focal lengths 2f are placed on same principal axis. If the object AB is placed between these mirrors, find the height ratio of the images of this object on two mirrors Hx/Hy=?

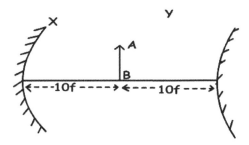

We use following equation to find height of the object;

$H_{image}/H_{object} = f/d_f$ where d_f is the distance between object and focal point of the mirror.

Distance between object and focal point of the mirror x is dfx=8f and distance between object and focal point of the mirror y is dfy=12f.

Let height of the object is h;

$H_{imageX}/h = 2f/8f$

$H_{imageX} = h/4$

$H_{imageY}/h = 2f/12f$

$H_{imageY} = h/6$

$H_{imageX}/H_{imageY} = 3/2$

Example: Ray (I) comes to optical system as shown in the figure given below. Ray reflects from the mirror placed inside the sphere in 1 direction. How much we must rotate mirror to make reflected ray follows path 2.

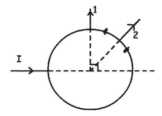

Since reflected ray rotates 45^0, mirror must rotate $22,5^0$

$45/2=22,5^0$

Example: An observer is placed at point O and an obstacle is placed at point A, which points shown in the figure below can be seen by observer?

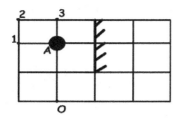

Picture, given below, shows which points are in the viewpoint of observer. Observer can only see image of object 3.

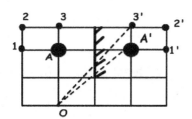

Example: As you can see in the picture given below, ray comes to the first mirror, reflects and hits second mirror. If this ray turns back on itself after reflecting from the second mirror, find the focal point of the second mirror.

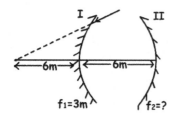

162

Ray passes from the point A after reflecting from first mirror. We do our calculations by taking point A object of second mirror.

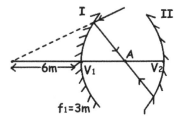

$1/f = 1/d_{object} - 1/6$

$1/3 + 1/6 = 1/d_{object}$

$d_{object} = 2m$

Point A is the center of the second mirror.

$V_1V_2 = 6m$

Center of the second mirror = 4m = 2f

f = 2m

Example: Relative refractive indexes of mediums are $n_{1,2}=2$, $n_{2,3}=1,5$. Find $n_{1,3}$.

$n_{1,2} = n_2/n_1 = 2$

$n_1 = n_2/2$

$n_{2,3} = n_3/n_2 = 1,5$

$n_3 = 1,5 n_2$

$n_{1,3} = n_3/n_1 = 1,5 n_2 / n_2/2 = 3$

Example: If the ray follows following path while passing from one medium to another, find the refractive index of medium.

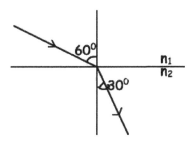

$n_{1,2} = n_2/n_1$

Using Snell's law

$n_1 \cdot \sin 60^0 = n_2 \cdot \sin 30^0$

$n_2/n_1 = \sin 60^0/\sin 30^0 = \sqrt{3}$

Example: Find the distance between point K and its image on plane mirror with respect to observer placed at point A.

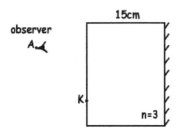

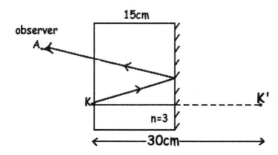

$h_{apparent} = n_{air}/n_{medium} \cdot h_{real}$

$h_{apparent} = 1/3 \cdot 30cm$

$h_{apparent} = 10cm$

Example: Path of the ray S is given below on spherical medium. Find the relation between refractive indexes of these mediums.

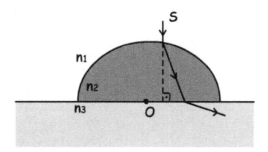

As you can see from the picture given below, angles of refracted rays are larger than the angles of incident rays, Thus, relation between the refractive indexed becomes;

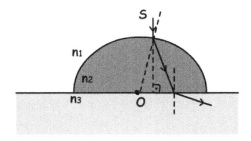

$n_1 > n_2 > n_3$

PROPERTIES OF MATTER

MATTER

Everything around us has mass and volume and they occupy space, and we called them as **matter**.

It can be in four sate, like solid, liquid, gas and plasma. We will talk about main properties of matter in this unit like, mass, volume, density, elasticity, inertia…Etc. You can classify matters with their physical or observable properties and chemical or unobservable properties, for example their smells, color, shapes give you an idea about it. On the contrary unobservable properties like conductivity of the matter cannot be understood from appearance or smells of the matter.

MASS

Mass is the quantity of the matter in a substance. We show mass with **m,** and unit of mass is gram (g) or kilogram (kg)**.** It is not the distinguishing property but common property of matters, because different matters may have equal mass.

INERTIA

Inertia is one of the properties of matter. It is the resistance of the matter to change its state of motion. An unbalanced force can only change the state of motion of the matter.

VOLUME

Volume is the space occupied by the matter. It is also common property of matter and does not help us in distinguishing them. We show it with **V** and unit used in SI system is m^3.

Volume formulas for some geometric shapes are given below.

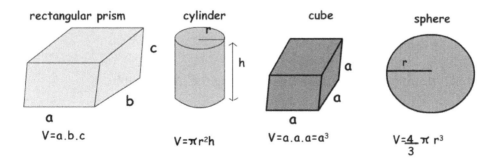

DENSITY

Density is the quantity of mass in a unit of volume. It is the distinguishing property of matter. Each matter has its own density. Representation of density is **d**; unit of it is **g/cm³**.

Formula of density;

$$d = \frac{mass}{volume} = \frac{m}{V}$$

We can show this relation with graphs also, look at the given graphs.

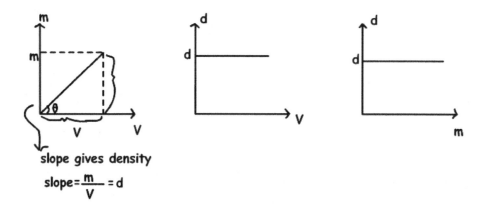

Be careful!! These graphs show the relations of mass, density and volume under constant temperature. Changes in the temperature change the values of volume and density. As you can see from the graphs increase in the volume and mass does not affect the value of density, because it is constant under constant temperature and characteristic property of matter.

Example: Find the densities of the given matters;

Matter having mass 100g and volume 50cm³ and matter having 125g and volume 100cm³.

Density=mass/volume=100g/50cm³=2g/cm³

Density=mass/volume=125g/100cm³=1,25g/cm³

Example: Find the density of the given matter using the given graph.

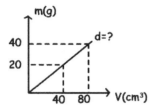

We find the slope of this graph which gives density.

slope = $\frac{40-20}{80-40}$ = 0,5

d=0,5g/cm³

Density of Mixtures

Mixtures include at least two different matters. If the matters are homogeneously mixed than we called them homogeneous mixtures, on the contrary if they do not mixed homogeneously we called them heterogeneous mixtures. Now we will learn how to calculate the density of homogeneous mixtures. Let me start with a mixture including two different matters.

Matter 1: has mass **m1** and volume **V1**

Matter 2: has mass **m2** and Volume **V2**

Density=mass/volume

If we have more than one substance;

Density of Mixtures = $\frac{\text{Total Mass}}{\text{Total Volume}} = \frac{m_1+m_2+m_3+\ldots+m_n}{V_1+V_2+V_3+\ldots+V_n}$

Example: Find the density of the mixture including three different matters having mass 25g, 35g and 65g, and volumes 10cm³, 25cm³, and 50cm³.

Density of Mixtures = $\frac{\text{Total Mass}}{\text{Total Volume}} = \frac{25g+35g+65g}{10cm^3+25cm^3+50cm^3}$

Density of Mixtures = $\frac{125g}{85cm^3}$ = 1,47g/cm³

Some Tricks for Special Cases in Density of Mixtures

If the volumes of the matters are equal in a given mixture then we calculate the density of mixture with following formula;

$$\text{If } V_1 = V_2 = V_3 = \ldots = V_n$$

$$d_{mixture} = \frac{d_1 + d_2 + d_3 + \ldots + d_n}{n}$$

If the masses of the matters are equal in a given mixture then we calculate the density of the mixture with following formula;

$$\text{If } m_1 = m_2$$

$$d_{mixture} = \frac{2 d_1 \cdot d_2}{d_1 + d_2}$$

Example: Graph given below shows the relation of mass vs. volume of two different matters A and B. If we take 40cm³ from A and 40cm³ from B, calculate the density of the mixture.

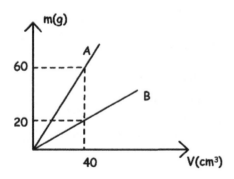

$$d_A = \frac{m_A}{V_A} = \frac{60g}{40cm^3} = 1{,}5 g/cm^3$$

$$d_B = \frac{m_B}{V_B} = \frac{20g}{40cm^3} = 0{,}5 g/cm^3$$

$$d_{mixture} = \frac{d_1 + d_2 + d_3 + \ldots + d_n}{n} = \frac{d_A + d_B}{2} = \frac{1{,}5 + 0{,}5}{2} = 1 g/cm^3$$

ELASTICITY

If an external force is applied to a material, it causes deformation in molecular structure of that material. By removing this force, material turns back to its original shapes; we call this process as **elasticity** of material. It is the ability of returning its original shape after removing the applied stress.

The Hooke's Law can explain elasticity. In other words, amount of compression or stretching is directly proportional to the applied force. However, until certain point under applied force materials regain their original shapes. This certain point is called **elastic limit** of that material. It is specific for each material and distinguishing property of solid matters. Look at the given pictures. You see the elasticity of the spring and rubber.

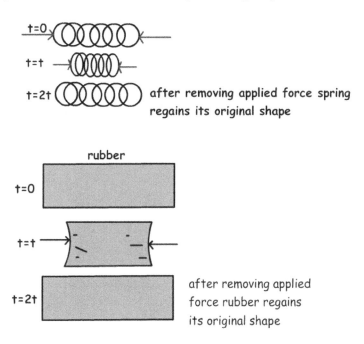

Forces are applied at time t. The external forces compress spring and rubber and after removing the forces they regain their original shape at time 2t. We can say that forces, applied to the spring and rubber, are below the elastic limit of these materials. If they are not below these limits rubber and spring are permanently distorted. Picture given below shows the example of this situation.

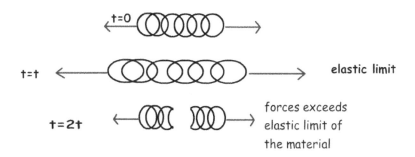

MORE EXAMPLES RELATED TO PROPERTIES OF MATTER

Example: Three different tubes are filled with same liquid having density d_1. If the rest of the tubes are filled with another liquid having density d_2, find the relation between the densities of the liquids in the tubes. ($d_2>d_1$)

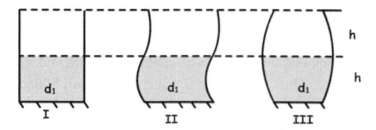

Heights of three tubes are equal, thus liquids mixed in the tubes have same volume. Densities of the liquids become;

$d_1=d_2=d_3$

Example: Mass vs. volume graph of two liquids is given below. If we take 160g X water and 40 cm³ from Y water, find the density of homogeneous mixture.

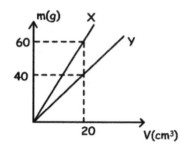

Using graph given above;

dx=80/20=4g/cm³

dy=40/20=2g/cm³

d$_{mixture}$=(mx+my)/(Vx+Vy)

d$_{mixture}$=(mx+Vy.dy)/mx/dx+Vy

d$_{mixture}$=(160+40.2)/160/4+40

d$_{mixture}$=3g/cm³

Example: We mix two different liquids having equal volumes and densities d_1=0,8 g/cm³ and d_2=d. If final density of the mixture is 1,2g/cm³, find the density of second liquid.

If the volumes of matters are equal we use following formula to find density of mixture;

$d_{mixture}=(d_1+d_2)/2$

$1,2=(0,8+d)/2$

$2,4-0,8=d$

$d=1,6 g/cm^3$

Example: Flow rates of two taps are equal and liquids have densities 3d and d. These two taps fill the half of the tank shown in the picture given below and rest of the tank is filled by the tap having liquid density d. Draw the density vs. time graph of the liquid in the tank.

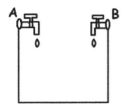

Since the flow rates of two taps are equal, density of the mixture in first half is;

$d_{mixture1}=3d+d/2=2d$

The tap having liquid density d fills rest of the tank. So, final density of the mixture becomes;

$d_{mixture2}=2d+d/2=1,5d$

Density vs. time graph of mixture is;

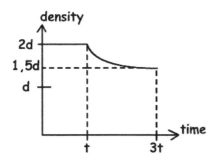

Example: A prism made of matter having density 5 g/cm³, has mass 600g. Find the volume of the space inside the prism.

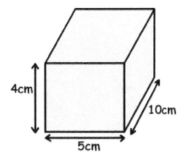

Exterior volume of the prism is;

$V_{exterior}$=4.5.10=200cm/³

Volume of the matter used for prism=600/5=120cm³

Thus;

Volume of the space inside prism is=200-120=80cm³

Example: There are three different tanks filled with water. If we fill rest of the tanks with liquid having density 0,8g/cm³, find the relation between the final densities of homogeneous mixtures.

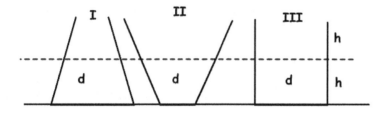

In the first tank, volume of water is larger than the volume of added liquid. Thus, density of mixture1 is closer to the density of water.

In the second tank, volume of water is smaller than the volume of added liquid. Thus, density of mixture2 is closer to the density of added liquid.

In the third tank, volume of water is equal to the volume of added liquid. Thus, density of mixture3 is ;

$d_{mixture3}$=(1+0,8)/2=0,9g/cm³

Relation between densities of mixtures;

$d_1>d_3>d_2$

Example: An object made of matter having density 6g/cm³ has mass 1500g. Find the volume of space inside the object.

m=V.d

1500=V.6

v=150cm/³

Object has volume 600 cm³, thus volume of the space is;

600-250=350cm³

Example: Table given below, shows the properties of three matters. Find whether these matters are same or not.

matter	volume (cm³)	mass (g)
X	8	24
Y	5	15
Z	8,5	25,5

If the densities of these matters X, Y and Z are equal at same temperature and pressure, we can say they are same matters.

$d_X=m_X/V_X=24/8=3g/cm^3$

$d_Y=m_Y/V_Y=15/5=3g/cm^3$

$d_Z=m_Z/V_Z=25,5/8,5=3g/cm^3$

X, Y and Z can be same matter.

Example: Which one/ones of the following properties is/are distinguishing property of matter.

Volume, Mass, Density, Elasticity, Melting Point

Volume and mass are common properties of matters. Density is distinguishing property of solid, liquid and gas form of matters. On the contrary, elasticity and melting point are distinguishing properties of solid matters.

HEAT TEMPERATURE AND EXPANSION

In this unit we will learn some concepts like heat, temperature, thermal expansion, thermal energy and phases of matter. Moreover, some misconceptions about heat and temperature will be explained. Since they make confusions in many students' mind, we give more importance on this subject. In daily life sometimes we use them interchangeably however; in physics they are totally different concepts. How we measure temperature of the substance, how much heat is required for melting the given ice, which matter expands much with the same amount of heat… We will try to answer all these questions in this section. Let's start with the definitions of heat and temperature.

TEMPERATURE

All matters are formed from atoms and molecules. In microscopic view we see that all particles in a matter are in random motion, they are vibrating, colliding randomly. We learned in previous sections that particle has kinetic energy if it moves. Thus, in an object all particles have kinetic energies because of their random motions. Temperature is the quantity, which is directly proportional to the average kinetic energy of the atoms of matter. Be careful, it is not energy, it just show the quantity of average kinetic energy of one atom or one molecule. In daily life we use some terms like hot, cold or warm. All these terms are used with respect to another reference matter. For example, you say that a glass of boiling water is hotter than the ice cream. Be careful, ice cream is our reference matter.

We measure temperature of matters with a device called **thermometer.** There are three types of thermometer, Celsius Thermometer, Fahrenheit Thermometer and Kelvin Thermometer. Look at the given picture to see how we scale thermometers. In Celsius thermometer, lower fixed point is 0 °C and upper fixed point is 100 °C, in Fahrenheit thermometer lower fixed point is determined as 32 °F and upper fixed point as 212 °F and finally, lower fixed point of Kelvin

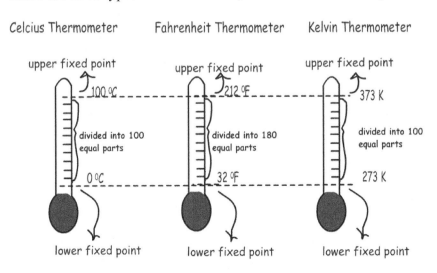

thermometer is 273 K and upper fixed point is 373 K. These temperatures are determined with considering the freezing point and boiling point of water.

We can convert the measurements of Celsius to Kelvin or Fahrenheit to Kelvin, Celsius by using following equations.

$$\frac{C}{100} = \frac{F-32}{180} = \frac{K-273}{100}$$

Example: Find the values of 30 °C in Fahrenheit and Kelvin thermometers.

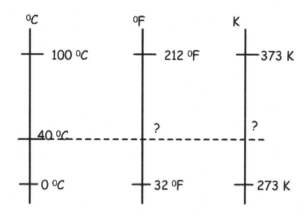

We use the equation given below to find corresponding values of 30 °C in Kelvin and Fahrenheit Thermometers

$$\frac{C}{100} = \frac{F-32}{180} = \frac{K-273}{100}$$

$$\frac{C}{100} = \frac{F-32}{180} \qquad \frac{C}{100} = \frac{K-273}{100}$$

$$\frac{40\ °C}{100} = \frac{F-32}{180} \qquad \frac{40\ °C}{100} = \frac{K-273}{100}$$

$$F = 104\ °F \qquad\qquad K = 313\ K$$

HEAT

Heat is a form of energy that flows from hotter substance to colder one.

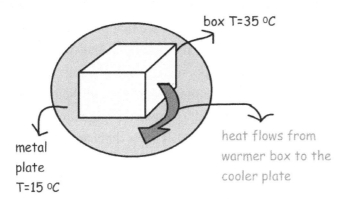

We mean by hotter and colder substance, substance having high temperature and low temperature, with respect to a reference matter. There must be a difference in temperatures of the substance to have heat or energy transfer. Heat is related to the quantity of matter also. If the object has big mass it also has big thermal energy and consequently amount of transferred energy increases. Since it is a type of energy we use Joule or Calories as unit of heat.

Differences between Heat and Temperature

In daily life most of us use these terms interchangeably. In this section we learn differences between them.

✦ Heat is a type of energy, but temperature is not energy.

✦ Heat depends on mass of the substance, however; temperature does not depend on the quantity of matter. For example, temperature of one glass of boiling water and one teapot of boiling water are equal to each other; on the contrary they have different heat since they have different masses.

✦ You can measure temperature directly with a device called thermometer but heat cannot be measured with a device directly. You should know the mass, temperature and specific heat capacity of that matter to find heat.

✦ If you give heat to a matter, you increase its temperature or change its phase.

Specific Heat Capacity

If you give same amount of heat to different type of matters you observe that changes in their temperatures are different. For instance, all you experience that given an equal amount of heat to metal spoons and wooden spoons, metal spoon has greater change in its temperature. Thus, most of the housewives use wooden or plastic spoons while cooking. These examples show that each matter has its own characteristics to absorb heat. We call this concept as **specific heat capacity** of the matters. It is the distinguishing property of matters. We show it with letter "**c**" and give the definition of it as, heat required to increase temperature of unit mass 1 °C. On the contrary, heat capacity of the system is defined as "heat required increasing the temperature of whole substance" and we show it with "**C**".

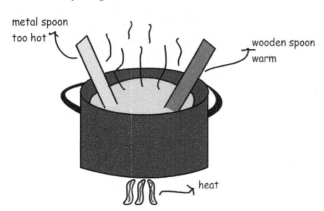

With the help of specific heat capacity and mass of the matter we can find the relation between heat and change in the temperature in the given formula below.

Q=m.c.ΔT

Where; Q is heat, m is mass, c is specific heat capacity and ΔT is the change in the temperature.

C=m.c where m is the mass of the substance and c is the specific heat of the matter.

HEAT TRANSFER

In previous section we have talked about heat. We said that, heat flows from the warmer objects to cooler ones. This process continues until the temperatures of the whole system become equal. Heat transfer occurs in three ways, convection, conduction and radiation. Now we look at the definitions and examples of these terms.

Conduction: When you give heat to an object the kinetic energy of atoms at that point increases and they move more rapidly. Molecules or atoms collide to each other randomly and during this collision they transfer some part of their energy. With the same way, all energy transferred to the end of the object until it reaches thermal balance. As you can see from the picture, atoms at the bottom of the object first gain energy, their kinetic energies increase, they start to move and vibrate rapidly and collide other atoms and transfer heat. Conduction is

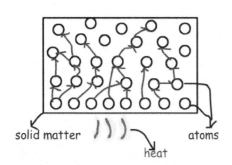

commonly seen in solids and a little bit in liquids. In conduction, energy transfer is slow with respect to convection and radiation. Metals are good conductors of heat and electricity.

Convection: In liquids and gases, molecular bonds are weak with respect to solids. When you heat liquids or gases, atoms or molecules, which gain energy, move upward, since their densities decrease with the increasing temperature.

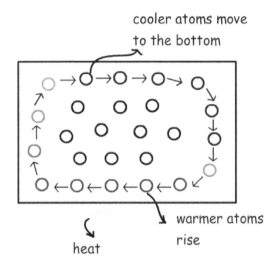

All heated atoms and molecules move upward and cooler ones sink to the bottom. This circulation continues until the system reaches thermal balance. This type of heat transfer does not work in solids because molecular bonds are not weak as in the case of fluids. Heat transfers is quick with respect to conduction.

Radiation: It is the final method of heat transfer. Different from conduction and convection, radiation does not need medium or particles to transfer heat. As it can be understood from the name, it is a type of electromagnetic wave and shows the properties of waves like having speed of light and traveling in a straight line. In addition to, it can travel also in vacuum just like sunlight.

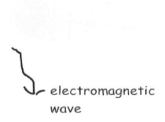

Radiation is a good method of transferring heat, in microwave ovens or some warming apparatus radiation is used as a method of heat transfer.

Calculation with Heat Transfer

Conservation of energy theorem is also applied to heat transfer. In an isolated system, given heat is always equal to taken heat or heat change in the system is equal to zero. If two objects having different temperatures are in contact, heat transfer starts between them. The amount of heat given is equal to the amount of heat taken. Object one has mass **m1**, temperature **t1** and specific heat capacity **c1**, object two has mass **m2**, temperature **t2** and specific heat capacity **c2**.

$$Q_{gained} = Q_{lost}$$
$$m_1 . c_1 . \Delta T_1 = m_2 . c_2 . \Delta T_2$$

Example: Find the final temperature of the mixture, if two cup of water having masses m1=150g and m2=250g and temperatures T1= 30 °C and T2=75 °C are mixed in an isolated system in which there is no heat lost. (cwater=1cal/g.°C)

$Q_{gained} = Q_{lost}$
$m_1 . c_1 . \Delta T_1 = m_2 . c_2 . \Delta T_2$
150g.1.(Tf−30 °C)=250g.1.(75 °C−Tf)
3Tf−90 °C=375 °C −5Tf
8Tf=465 °C
Tf=58,1 °C is the final temperature of the water mixture

Example: Temperature of the iron block decreases from 85 °C to 25 °C. If the mass of the block is 1,2kg, calculate the heat lost by the block. (c_{iron}=0.115cal/g.°C)

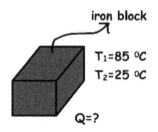

$Q_{lost} = m_{iron} . c_{iron} . \Delta T$
Q_{lost}=1200g.0,115cal/g.°C.(85 °C−25 °C)
Q_{lost}=8280 Calorie is the heat lost by the block

Example: The graph given below shows the relation between given heat and change in the temperatures of the three matters having same masses. Compare the specific heat capacities of these matters.

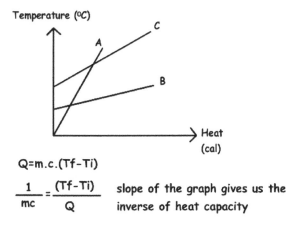

$Q = m \cdot c \cdot (T_f - T_i)$

$\dfrac{1}{mc} = \dfrac{(T_f - T_i)}{Q}$ slope of the graph gives us the inverse of heat capacity

Since the masses of these matters are equal, B has the greatest specific heat capacity because, with the same amount of heat, change in the temperature of the B is lower than the other two matters. Moreover, A has the minimum specific heat capacity, because the change in its temperature with the same amount of heat is larger than the others. Finally, specific heat capacity of the C is between A and B. Thus;

$c_B > c_C > c_A$

CHANGE OF PHASE/STATE

Matters can be in four states like solid, liquid, gas and plasma. Distance between molecules or atoms of the matter shows its state or phase. Temperature and pressure are the only factors that affect the phases of matter. Under constant pressure, when you heat matter, its speed of motion increases and as a result distance between atoms or molecules becomes larger.

If you give heat to a solid substance, its temperature increases up to a specific point and after this point temperature of it becomes constant and it starts to change its phase from solid to liquid. Another example that all you experience in daily life, when you heat water it boils and if you continue to give heat it starts to evaporate. In this section we will learn these changes in the phases of substances and learn how to calculate necessary heat to change the states of them.

Melting and Freezing

If solid matters gain enough heat they change state solid to liquid. Heat is a form of energy and in this situation it is used for the break the bonds of the atoms and molecules. Heated atoms and molecules vibrate more quickly and break their bonds. We call this process **melting** changing state solid to liquid. Inverse of melting is called **freezing,** changing state liquid to solid, in which atoms and molecules lost heat and come together, their motion slows down and distance between them decreases.

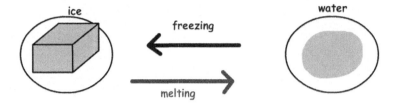

Look at the given graph, which shows melting of the ice.

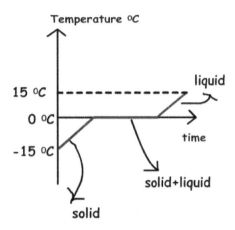

This is a phase of change of water from solid to liquid. As you can see at the beginning ice is at -15 °C, we give heat and its temperature becomes 0 °C which is the melting point of ice. During melting process temperature of the ice-water mixture does not change. After all the mass of ice is melted its temperature starts to rise.

Every solid matter has its own melting point; we can say that melting point is a distinguishing property of solids. Inverse of this process is called freezing in which liquid lost heat and change phase liquid to solid. Freezing point and melting point are the same for same matter and it is also distinguishing property of matter. We find the heat necessary for melting the solid substance with following formula;

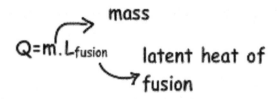

L_{fusion}, like specific heat, it shows how much heat you should give for melting/freezing unit of mass. For example, $3,3 \times 10^5$ joule/kg is the latent heat of fusion for ice and liquid.

Example: Find the amount of heat for melting ice having mass 1,3kg at -10 °C? (L_{fusion} =3,3×10⁵ joule/kg c_{ice}=2,2X10³j/kg.°C)

We first increase the temperature of the ice from -10 °C to 0 °C (melting point).

Q_1 = m. c (Tf – Ti)

Q_1 = 1.3kg. 2.2x 10³j/kg.⁰C (0-(-10))

Q_1 = 28600 joule

Now we find heat necessary for melting ice at 0 °C.

Q_2 = m . L_{fusion}

Q_2 = 1.3kg. 3.3x10⁵

Q_2 = 429000 joule

Total heat required for whole process is;

Q_T = Q_1 + Q_2 = 28600joule + 429000joule

Q_T = 457600j joule

Effects of Pressure and Impurity on Freezing and Melting Point

Pressure is the force exerting on the surface perpendicularly. Thus, it helps to keep particles together. If volume of the matter increases after melting, pressure decreases the melting point. On the contrary, if the volume of the substance decreases after melting, pressure increases the melting point of the matter.

For example, when you walk on the snowy road you observe that snow under your feet melt later than around, because you exert pressure on it with your feet. Ices melting at 0 °C can be melt at -3 °C with the applied pressure on it. Impurity like pressure affects the latent heat of fusion. For instance, salty water freezes under 0 °C.

Boiling Evaporation and Condensation

Evaporation is the change of phase from liquid to gas. Evaporation occurs only at the surface of the water and at every temperature. However, evaporation is directly proportional to the temperature, increasing in the temperature increasing in the rate of evaporation. Inverse of this process is called condensation in which; gas molecules/atoms lost heat and change phase from gas to liquid.

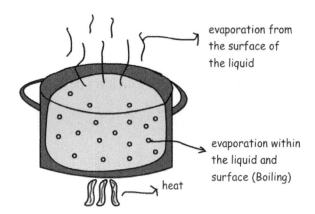

As in the case of melting, when you give heat to liquid, at one certain point its temperature does not change. Gained heat spent on breaking the bonds between molecules and atoms. At this temperature, vapor pressure of the liquid is equal to the pressure of surrounding. During this process evaporation occurs in everywhere of the liquid, which is called **boiling**. Boiling point is a distinguishing property of liquids; each matter has its own boiling point. For example, water boils at 100 ºC in atmospheric pressure. We use the following formula to find required heat to boil liquid matter.

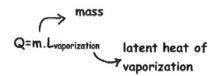

Where; m is the mass of the liquid matter and $L_{vaporization}$ is the latent heat of vaporization that shows the necessary heat to evaporate unit of mass. For example, you should give $2,3 \times 10^6$ joule heat to change the phase of water from liquid to gas.

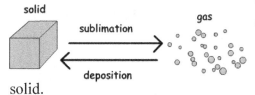

Sublimation is the change of state from solid to gas. Some of the solid matters change their states directly to the gas with the gained heat. For example, dry ice (frozen CO_2) sublimate when heat is given. Inverse of this process is called deposition, in which gas matters lost heat and change their phase to solid.

Example: Find the amount of heat for evaporating 2,8kg of water at 45 ºC? ($L_{vaporization}$ =$2,3 \times 10^6$ joule/kg c_{water}=4190j/kg.ºC)

First we should increase the temperature up to 100 ºC.

$Q_1 = m.c.(T_f - T_i)$

$Q_1 = 2.8\text{kg} \cdot 4190 \text{j/kg.}^0\text{C} \ (100 - 45)$

$Q_1 = 645260$ joule

Now we have liquid at 100 ^{0}C and we can apply latent heat of vaporization formula.

$Q_2 = m.L_{vaporization} = 2.8\text{kg} \cdot 2.3 \times 10^6 \text{j/kg.}^0\text{C}$

$Q_2 = 6.44 \times 10^6$ joule

Total heat required = $Q_1 + Q_2$ = 7085260 joule

Effects of Pressure and Impurity on Boiling Point

Boiling occurs only when the vapor pressure of liquid and pressure of outside equals to each other. If the pressure of outside increases then the boiling point of the liquid also increases. On the contrary, if the pressure of the outside decreases, then boiling point of the liquid also decreases. For example, at the top of a mountain atmospheric pressure is lower than the atmospheric pressure of the sea level. In addition to this, impurity of the liquid matter also affects the boiling point of that matter.

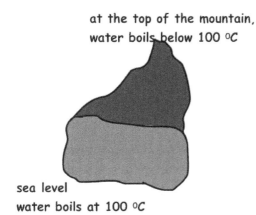

at the top of the mountain, water boils below 100 ºC

sea level
water boils at 100 ºC

For instance, if you mix water with a salt or sugar, you increase the boiling point of the water.

Example: Graph given below shows the relation of temperature and gained heat on different matters. Which ones of them are possible?

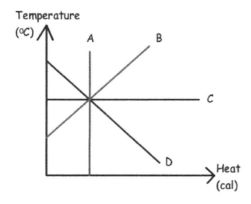

Line of A shows that matter gains no heat but its temperature increases. Such a relation between heat and temperature of the matter is not possible.

Line B shows that, temperature of the substance increases with the gained heat. It is possible.

Line C shows that, matter gains heat but its temperature stays constant. This is also possible; phase of matter C can be changing.

Line D says that, matter gains heat however, its temperature decreases. This situation is not possible.

PHASE TRANSITION OF WATER

This graph shows the phase diagram of water. At the beginning we have ice at -20 °C. Ice

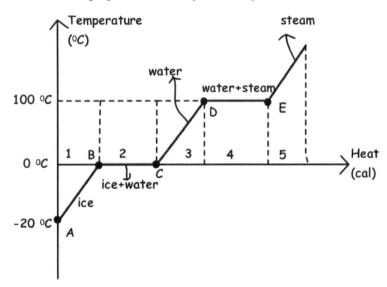

gain heat in the interval of points A and B, and its temperature becomes 0 °C that is the melting point of it. We have only ice in the 1st region. As you can see between the points B and C, temperature of the mass does not change, because its state is changing in this interval. Gained heat is spent on t breaking the bonds of molecules. 2nd region includes both water and ice. After melting process completed, in the 3rd region there is only water and temperature of water starts to increase. When the temperature of the water becomes 100 °C, it starts to boil and evaporation of it speeds up. In region 4 our mass exists in two state, water and steam. After completion of evaporation, all water converted to the steam and in region 5 we have only vapor of water.

This graph shows the condensation of water vapor, which lost heat. As it seen from the graph, steam lost heat and its temperature decreases at 100 °C, at this temperature it condensate and becomes water, after heat lost it reaches at temperature 0 °C and starts to freeze. Finally, if it continues to lost heat, their temperature also continues to decrease.

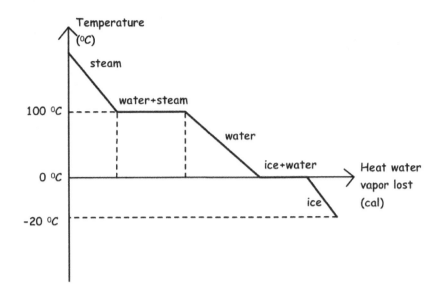

COMMON MISCONCEPTIONS

✦ If a matter gains heat then its temperature can be increase or it can change phase. On the contrary, if a matter lost heat, its temperature can be decrease or it can change phase.

✦ Change in temperature of the matter causes change in the internal kinetic energy of the matter, but change of phase causes change of internal potential energy of the matter.

✦ Time of melting and boiling depend on the amount of matter.

✦ There is no heat transfer between matters having same temperatures.

✦ Except from water, densities of all matters are inversely proportional to the temperature.

THERMAL EXPANSION AND CONTRACTION

Most of the matters, without some exceptions, expand with the increasing temperature. When you give heat to matters; speed of its particles increase and distance between them also increases which results in the increase of the volumes of matters.

All expansions occur in volume of the substance however; sometimes some of the dimensions of them expand more with respect to others. In this case we neglect the less expanded ones and assume expansion like **linear expansion** in long materials. Moreover, we take the expansion of plate as **area expansion** and finally we take the expansion in three dimensions as **volume expansion**.

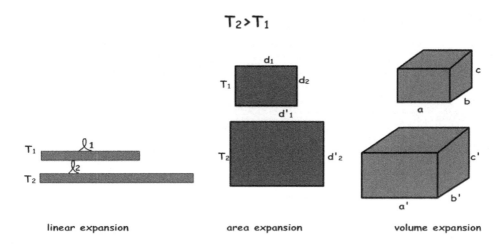

linear expansion   area expansion   volume expansion

Inverse of the expansion is called **contraction**, generally when matters lost heat and their temperatures decrease they contract. Now we will learn which factors affect expansion.

If the initial volumes, areas or lengths of the matters are big enough their expansions with the same temperature are also big. In other words, expansion or contraction is linearly proportional to the initial volume of the matter.

Different matters have different atomic structure, thus distances between the atoms are also different. They give different reactions to the same amount of temperature changes. So, another factor effecting expansion is type of matter.

Final factor that affects expansion is the amount of change in temperature. Larger the change in temperature results in larger the change in the volume of matter.

We get following formula from the explanations given above;

$\Delta V = V_0 \cdot \alpha \cdot \Delta T$

Where; ΔV is the change in the volume, α is the coefficient of thermal expansion and ΔT is the change in the temperature of the matter.

α= Coefficient of thermal expansion is equal to the change in the volume of a unit of mass under 1^0C change in temperature.

Expansion in Solid Matters

We will examine this subject under three titles, linear expansion, area expansion and volume expansion.

Linear Expansion:

Picture given below shows the linear expansion of metal rod. When it is heated, its length increases.

linear expansion

T_1 ▭ ℓ_1

$T_2 > T_1$

T_2 ▭ ℓ_2

Our formula for linear expansion is;

ΔL=L0.α. ΔT

Where; ΔL is the amount of change in the length of the rod, L_0 is the initial length of the road, α is the coefficient of linear expansion and ΔT is the change in the temperature of the matter.

Example: There are three same metal rods having same length and thickness. If the temperatures of them are given like; T, 2T and 3T find the relations of final lengths of the rods. (Rods are in contact)

We find the final temperatures of the system by the formula;

$T_{final}=T_1+T_2+T_3/3=6T/3=2T$

Since the temperature of the first rod increase, its final length also increases. Temperature of the second rod stays same, thus there won't be change in the length of this rod. Finally, temperature of the third rod decreases, thus it contract and final length of it decreases with respect to initial length. As a result relation of the final lengths of the rods;

$L_1 > L_2 > L_3$

Area Expansion:

When plate given below is heated, it expands in two dimensions X and Y.

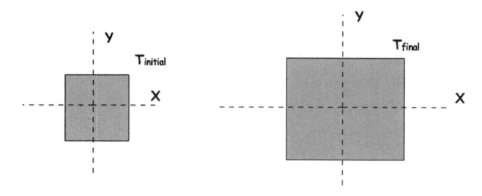

We find the area expansion with the given formula;

$\Delta S = S_0 \cdot 2\alpha \cdot \Delta T$

Where; ΔS is the amount of change in the area of the plate, S_0 is the initial area of the plate, 2α is the coefficient of area expansion and ΔT is the change in the temperature of the matter.

Example: We cut a circular piece from the rectangular plate. Which ones of the processes given below can help us in passing through the circular piece from the hole?

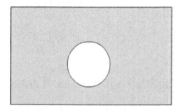

I. Increasing the temperatures of rectangular plate and circular piece

II. Decreasing the temperature of the circular piece

III. Decreasing the temperatures of the rectangular plate and circular piece

I. If we increase the temperatures of the plate and circular piece, expansion of the hole and the circular piece will be the same. Thus, this option can help us.

II. If we decrease the temperature of the circular piece, it contracts and hole becomes larger than the piece. This option can also help us.

III. If we decrease the temperatures of the plate and circular piece, hole and circular piece contract in same size. This process can also help us.

Volume Expansion:

If the objects expand in volume with the gained heat, we call this volume expansion and find it with the following formula;

$\Delta V = V_0 \cdot 3\alpha \cdot \Delta T$

Where; ΔV is the amount of change in the volume of the cube, V_0 is the initial volume of the cube, 3α is the coefficient of volume expansion and ΔT is the change in the temperature of the matter.

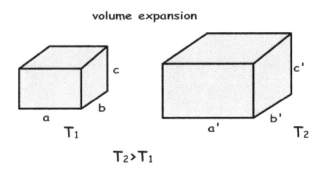

MORE EXAMPLES RELATED TO HEAT TEMPERATURE AND THERMAL EXPANSION

Example: Two thermometers X shows boiling point of water 220X and freezing point of water 20X and Y shows boiling point of water 120 Y and freezing point of water -40Y. If thermometer X shows 100X then find value of thermometer Y shows.

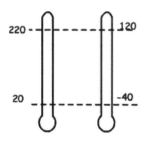

(X-20)/200=(Y-(-40))/160

(X-20)/20=(Y+40)/16

Y=24^{0}Y

Example: Two matters have specific heat capacities c and 2c. If we give Q and 4Q heat to these matters, changes in the temperatures of them become equal. If the matter A has mass m, find the mass of matter B in terms of m.

Heat gained, lost by the matters is found with following formula;

Q=m.c.ΔT

Heat gained by A and B;

Q=m$_A$.c.ΔT

4Q=m$_B$.2c.ΔT

m$_B$=2m$_A$

m$_B$=2m

Example: If A and B matters have equal masses find the ratio of c_A/c_B.

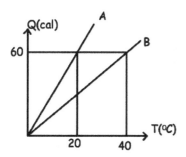

$Q = m.c.\Delta T$

$30 = m.c_A.(10-0)$

$30 = m.c_B(20-0)$

$c_A/c_B = 2$

Example: Table given below shows initial lengths, changes in temperatures and changes in the length of 3 rods. Find whether these rods are made of same matters or not.

matter	L₀	ΔT	ΔL
A	2L	ΔT	ΔL
B	3L	2ΔT	3ΔL
C	4L	ΔT	ΔL

$\Delta L = L_0.\alpha.\Delta T$

$\alpha_A = \Delta L / 2L.\Delta T$

$\alpha_B = 3.\Delta L / 3L.2\Delta T$

$\alpha_C = \Delta L / 4L.\Delta T$

$\alpha_A = \alpha_B > \alpha_C$

Thus, A and B can be same matter but C is different from them.

Example: Three cylinders made of same matter and at same temperature, are placed on a platform. Same amount of heat given to the cylinders make same amount of change in their temperature. Find the relation between changes in the lengths of these cylinders.

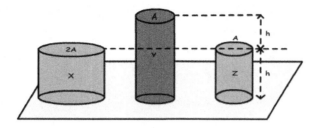

If we say volume of X is V, then volume of Y is also V and Z is V/2. There is a linear relation between volume and mass; we use this relation in solution of the problem.

$Q=m.c.\Delta T_X$, $\Delta T_X=T$

$Q=m.c.\Delta T_Y$, $\Delta T_Y=T$

$Q=m/2.c.\Delta T_Z$, $\Delta T_Z=2T$

Changes in the lengths of cylinders;

$\Delta L_X=h.\alpha.T=\Delta L$

$\Delta L_Y=2h.\alpha.T=2\Delta L$

$\Delta L_Z=h.\alpha.2T=2\Delta L$

$\Delta L_Y=\Delta L_Z>\Delta L_X$

Example: If Celsius thermometer shows the temperature of air 30^0C, find the temperature of air in Fahrenheit and Kelvin thermometer.

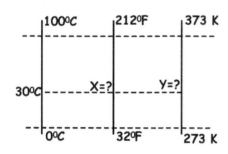

$T(K)=T+273$

$T=30+273=303\ K$

$C/100=(F-32)/180$

$30/100=(F-32)/180$

$F=86^0F$

Example: Find heat required to make 5g ices at -20°C to water at 30°C. (c_{ice}=0,5cal/g.°C, L_{ice}=80cal/g, c_{water}=1cal/g.°C)

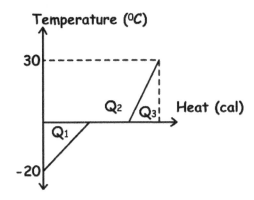

Heat required to make ice at -20°C to ice at 0°C ;

Q_1=m.c_{ice}.ΔT=5.0,5.20

Q_1=50 cal

Heat required making it melt;

Q_2=m.L_{ice}=5.80 = 400 cal

Heat to make it water at 30°C;

Q_3=m.c_{water}.ΔT=5.1.30 = 150 cal

Q_{total}=Q_1+Q_2+Q_3=50+400+150 = 600 cal

Example: Two taps fill the water tank with different flow rates. Tap A fills the tank in 1 hour and tap B fills the tank in 3 hour If we open two taps together, find the final temperature of the water in the tank.

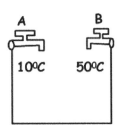

Flow rates of taps;

V_A=3V_B

3m.c.(T-10)=m.c(50-T)

T=20°C

Example: When we decrease the temperatures of the rods ΔT, relation between the final lengths of rods becomes; $L_1<L_2<L_3$. Find the relation between $α_A$, $α_B$ and $α_C$.

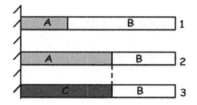

Since change in length of the rod 1 is larger than the rod 2;

$α_A<α_B$

Since change in length of the rod 2 is larger than the rod 3;

$α_C<α_A$

$α_C<α_A<α_B$

Example: Length vs. temperature graph of A, B and C is given below. Find the relation between $λ_A$, $λ_B$ and $λ_C$.

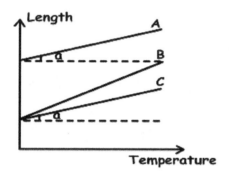

Slope of the graph=$ΔL/ΔT=L_0.λ$

Initial length of A is larger than C but slopes of them are equal, so;

$λ_C>λ_A$

B and C have same length but slopes of them are different.

$λ_B>λ_C$

$λ_B>λ_C>λ_A$

ELECTROSTATICS

Scientist found that if you rub an ebonite rod into silk you observe that rod pulls the paper pieces. Or in winter when you put off your pullover, your hair will be charged and move. We first examine the structure of atom to understand electricity better. Experiments done show that there are three types of particle in the atom.

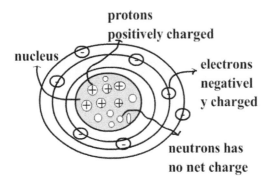

Structure of the Atom

Two of them are placed at the center (nucleus) of the atom, which we called proton (p) and neutron (n). Proton has positive charges "+" and neutron has no net charge. Third particle is called electron (e) and they are placed at the orbits of the atom. They are negatively charged "-". Electrons can move but proton and neutron of the atom are stationary. We show charge with "q" or "Q" and smallest unit charge is 1.6021×10^{-19} Coulomb. One electron and a proton have same amount of charge.

Positively Charged Particles

In this type of particles, numbers of positive ions are larger than the numbers of negative ions. In other words numbers of protons are larger than the number of electrons.

$p^+ > e^-$

To neutralize positively charged particles, electrons from the surroundings come to this particle until the number of protons and electrons become equal. Do not forget protons cannot move!

Negatively Charged Particles

In this type of particles, numbers of negative ions are larger than the numbers of positive ions. In other words numbers of electrons are larger than the number of protons.

$e^+ > p^-$

To neutralize negatively charged particles, since protons cannot move and cannot come to negatively charged particles, electrons moves to the ground or any other particle around itself.

Neutral Particles

These types of particles include equal numbers of protons and electrons. Be careful, they have protons, neutrons and electrons however, numbers of "+" ions are equal to the numbers of "-" ions.

$e^+ = p^-$

Conductors

Some of the matters have lots of free electrons to move. It is easy for electrons to flow from these materials. Metals are good conductors. Gold, copper, human bodies, acid, base and salt solutions are example of conductors.

Insulators

These types of materials do not let electrons flow. Bonds of the electrons in the insulators are tighter than the conductors thus, they cannot move easily. Glass, ebonite, plastic, wood, air are some of the examples of insulators.

Atoms having same charge repel each other and atoms having opposite charges attract each other.

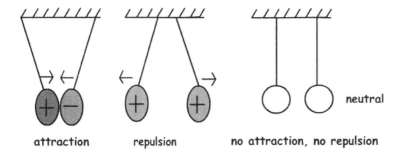

Example: Charged spheres A, B and C behave like as shown in the picture under the effect of charged rod D and E. If C is positively charged, find the signs of the other spheres and rods.

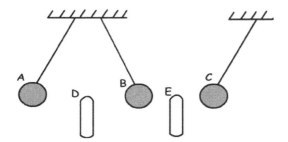

We learned that opposite charges attract each other and same charges repel each other. Using this explanation we can say that, if the sign of the C is "+" than rod E must be "-" since it attracts C. B must be "+" since E also attract B. Rod D repels the B so, we say that D must have same sign with B "+", and finally D also repels A, thus A is also "+".

A (+), D (+), B (+), E (-), C (+)

TYPES OF CHARGING

Charging means gaining or losing electron. Matters can be charged with three ways, charging by friction, charging by contact and charging by induction.

Charging by Friction

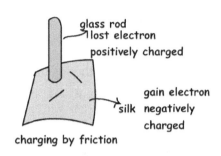

charging by friction

When you rub one material to another, they are charged by friction. Material losing electron is positively charged and material that gain electron is negatively charged. Amount of gained and lost electron is equal to each other. In other words, we can say that charges of the system are conserved. When you rub glass rod to a silk, glass lose electron and positively charged and silk gain electron and negatively charged.

Charging by Conduction

Thee are equal numbers of electrons and protons in a neutral matter. If something changes this balance we can say it is charged. Look at the following picture;

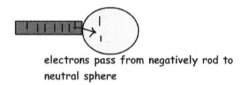

electrons pass from negatively rod to neutral sphere

In this picture, negatively charged rod touches to the neutral sphere and some of the electrons pass to the sphere. As a result neutral sphere is charged by contact. If the rod is positively charged, then some of the electrons of sphere pass to the rod and when we separate them, sphere becomes positively charged. Picture given below shows the flow of electrons from sphere to the rod.

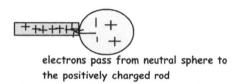

electrons pass from neutral sphere to the positively charged rod

1. When charged object touches to a neutral object, they both have same charge.

2. When two charged matter touch each other, total charge of the system is conserved and they share the total charge according to their capacities. If they have same amount of different charges, when we touch one another they become neutral. If the amount of charges is different then, after flow of charge they are both negatively or positively charged. Having opposite charges after contact is impossible.

3. If the touching objects are spheres, they share the total charge according to their radius, because their capacities are directly proportional to their radius. When the spheres are identical then they share total charge equally.

4. We can find the charge per radius by the following equation;

$$q = \frac{q_{total}}{r_{total}} = \frac{q_1+q_2+q_3+\ldots}{r_1+r_2+r_3+\ldots}$$

Then we use following equations to find the charge per sphere;

$$q_1 = \left(\frac{q_1+q_2}{r_1+r_2}\right) \cdot r_1 \quad \text{and} \quad q_2 = \left(\frac{q_1+q_2}{r_1+r_2}\right) \cdot r_2$$

Example: Charged spheres having radius 3r, 2r and r have different charges. If we touch three spheres to each other, find the final charges of the spheres.

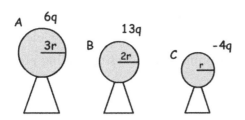

$$q_r = \frac{q_{total}}{r_{total}} = \frac{6q+13q+(-4q)}{3r+2r+r} = \frac{15q}{6r}$$

$$q_A = \left(\frac{15q}{6r}\right) \cdot r_A = \frac{15q}{6r} \cdot 3r = 7,5q$$

$$q_B = \left(\frac{15q}{6r}\right) \cdot r_B = \frac{15q}{6r} \cdot 2r = 5q$$

$$q_C = \left(\frac{15q}{6r}\right) \cdot r_C = \frac{15q}{6r} \cdot r = 2,5q$$

Charging by Induction

We can also charge conductors without contact. Examine the given picture, it shows this type of charging.

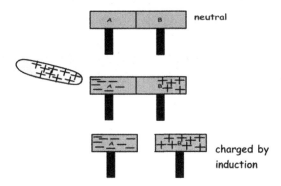

A and B conductors are neutral at the beginning. When we put a positively charged plate near them, it attracts the electrons in the conductors. Electrons move to the left part and protons stays. Thus, when we separate plates A and B they are charged by induction, A is negatively charged and B is positively charged. Be careful, there is no contact; they are charged only by induction.

GROUNDING

Grounding means making objects neutral. If it is negatively charged than taking its electrons or if it is positively charged than make it gain electrons. Universe has excess amount of electrical charges, electrons and protons. This huge resource makes all charged particles neutral we call this process as **grounding.** Look at the given example that shows grounding process of positively charged matters and negatively charged matters.

First sphere is negatively charged. When we connect it to ground excess amount of electrons flow to the ground and sphere becomes neutral. In the second sphere, it is positively charged. As you learned before, protons cannot move, thus electrons with the same amount of protons must come to the sphere to make it neutral. Sphere attracts the electrons from the ground and becomes neutral.

In the first picture, there is a neutral sphere. We want to charge it positively. First, we put a negatively charged rod near the sphere that repels the electrons right side of the sphere and protons stay at left side of the sphere. After separation of the electrons and protons we ground the sphere and make electrons flow through the earth. Finally, after the completion of electron flows we break the grounding and take the negatively rod away. We negatively charge the sphere with the help of grounding and induction.

Electroscope

It is a device that is used for detecting whether an object is charged or uncharged. It is also determine the type of charge.

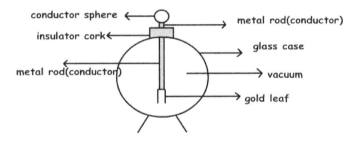

As you can see from the picture there is a metal (conductor sphere) and a metal rod attached to this sphere. Gold leaves are at the bottom of the rod. Electroscopes are placed in a glass case to diminish the effects of wind and ions in the air. This picture shows the neutral electroscope. If we charge it, because of the fact that same charges repel each other, leaves of the electroscope rises. Amount of the distance between the leaves depends on the amount of charges electroscope has. We analyze the different type of charging electroscope below.

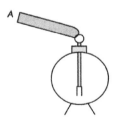

In this example, conductor **A** touches to the sphere of the neutral electroscope, however leaves of it stay closed. We understand that A is neutral; if it were not neutral leaves would rise.

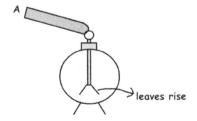

Here, as you can see leaves of the electroscope rise with the coming charges from the conductor A. We do not know the type of charge but we can say that A is charged, not neutral. Below examples show how we can determine the types of charges.

To understand the type of charge of conductor, we should bring it closer to the charged electroscope.

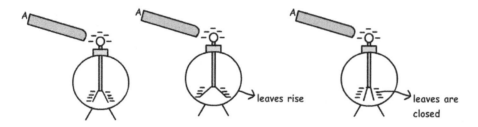

Picture given above shows the behavior of the leaves of negatively charged electroscope under the effect of charged rod. In the first picture, we bring rod closer to the sphere of negatively charged electroscope to observe motion of leaves. If the rod A is negatively charged, then same charges repel each other, electrons at the sphere are repelled to the end of the electroscope (to the leaves), and leaves rise with excess amount of electrons. Second picture shows this assumption. In the third picture, we assume that if the rod is positively charged, then rod attracts some of the electrons to the sphere and amount of electron on the leaves decreases. This decrease in the number of electrons makes them close. To sum up; when you bring charged object closer to the charged electroscope

- If the leaves rise then both the rod and electroscope have same charge.

- But if the leaves are closed than charges of electroscope and rod are opposite.

Now we will examine the behavior of leaves when we touch charged object to the charged electroscope. There are different situations in this case, let's look them one by one (we assume that capacities of electroscope and rod are same).

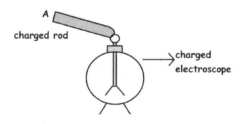

1. If A has +q charge and electroscope has –q charge, total charge becomes zero and electroscope and rode becomes neutral and leaves are closed.

2. If A has +3q charge and electroscope has -4q, total charge becomes –q. Electroscope and rod share –q charge, thus, leaves of electroscope are closed a little bit. Be careful, they are not closed totally.

3. If A has +4q charge and electroscope has -2q charge, total charge of the system becomes +2q. Electroscope and rod share +2q charge, electroscope becomes positively charged. During this process, leaves of electroscope are closed and after completion of charge sharing they rise again.

Example: Positively charged electroscope and Y-Z plates are placed in the figure given below. When we put T conductor sphere next to the Y-Z plates, distance between the leaves of electroscope increases. Find the type of charges of the Y, Z and T.

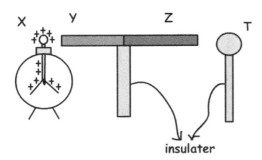

Y-Z plates are charged by induction when T conductor placed next to them. To increase the distance between the leaves of the electroscope, it must be under the effect of same charge. So, Y must be +, to make leaves rise more. Moreover, if Y is +, then Z must be -, because opposite charges repel each other (electroscope attracts the electrons of the Y-Z plates to the left). If Z is "–" then T is automatically becomes "+" to attract them.

Example: Negatively charged electroscope and X-Y plates are placed like in the figure given below. If the distances between electroscope and conductor plates are equals; find the types of charges of plates for given situations.

A. If there is no change in the distance between the leaves

B. If the leaves are closed a little bit

C. If the leaves rise a little bit

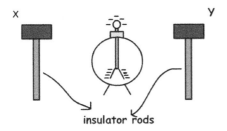

A. If there is no change in the behavior of the leaves, then effect of one plate is canceled by other plate. This can be possible under the circumstances, amount of charge of X is equal to amount of charge of Y and one of them must be negatively charged and other one must be positively charged.

B. If the leaves are closed a little bit, then some of the electrons on the leaves must be move to the top of the electroscope. This can be possible with following ways;

I. Both of the plates can be positively charged and have different amounts of charges. They both attract electrons of the leaves.

II. They can have different charges; however, amount of positive charge must be larger than the negative charge. Since, there must be net positive charge to attract electrons from the leaves.

C. In the last situation, if the leaves rise a little bit, then some of the electrons from the top move to the leaves. This can be possible with following ways;

I. Both of the plates can be negatively charged and have different amounts of charges. They repel electrons from top of the electroscope to the leaves.

II. They can have different types of charges, but amount of negative charge must be larger than the amount of positive charge to repel electrons from top of the electroscope to the leaves.

Special Cases

Look at the given picture; when we put a charged conductor sphere into the neutral hollow, it is charged by induction.

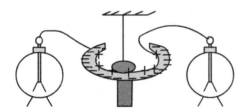

Negative charges are repelled to the outer part and positive charges stay at inner part. We connect neutral electroscopes to the inner and outer part of the hollow. Leaves of neutral electroscopes rise that shows us it is charged by induction.

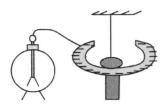
When we touch negatively charged sphere into the inner part of hollow, we observe that sphere becomes neutral and electrons are repelled to the outer part of hollow. If we connect neutral electroscope to the outer part of the hollow, leaves of it rises, since outer part of hollow is negatively charged by contact. Be careful, inner part of hollow and sphere are becomes neutral.

Picture given below shows the charged conductor. As you can see, same charges repel each other and they are located as far as possible to each other. Density of charges at left and right points of the conductor is larger than the middle part.

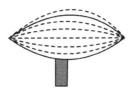

Charged plate attracts the neutral sphere. Here we see the example of charging by induction. Electrons move to the right side of the sphere and attraction force is larger than the repulsion, thus sphere moves toward to the plate. If both plate and sphere are charged with same types of charge they repel each other.

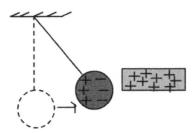

In the next section we will learn electrostatics forces between the charges and find the amount of forces.

ELECTRICAL FORCES COULOMB'S LAW

In the previous sections we learned that same charges repel each other and opposite charges attract each other. Experiments done on this subject show that this force is depends on the distance between the charges and amount of charges.

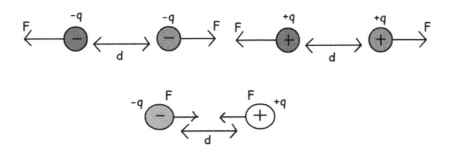

If the distance between the charged object increase, then amount of attraction or repulsion decrease. On the contrary if the distance between them decreases then amount of force increases. Moreover, electrical forces are directly proportional to the amount of net charge. Coulomb made some experiments and find following equation of electrical forces.

$$F = k \frac{q_1 \cdot q_2}{d^2}$$

where; q_1 is the amount of charge on object1
q_2 is the amount of charge on object2
k is the proportionality constant
d is the distance between two objects

- If the objects have same type of charge then the force is repulsive, if they have opposite charges then force is attractive.

- Repulsive or attractive electrical forces is equal in magnitude but opposite in direction, it does not depend on the magnitude

$\mathbf{F_1 = -F_2}$

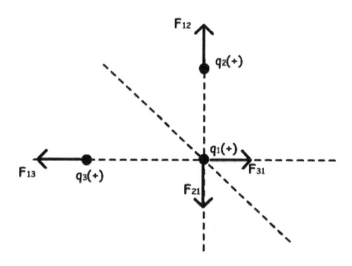

Picture given above shows the forces acting on the charges. F_{21} is the force of repulsion of F_2 on F_1; F_{12} is the force of F_1 on F_2. As I said before;

$F_{12} = -F_{21}$

Example: If the q_2 charge does not move, find the charge of q_3 in terms of q?

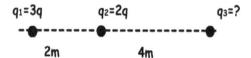

q₁=3q F₃₂   q₂=2q F₁₂   q₃=?
●←――――●――――→-----●
   2m   4m

$F_{12}=F_{32}$ they are equal in magnitude and opposite in direction

$$F_{12}=k\frac{q_1.q_2}{d_1^2} \qquad F_{32}=k\frac{q_3.q_2}{d_2^2}$$

$$F_{12}=k\frac{3q.2q}{2^2} \qquad F_{32}=k\frac{q_3.2q}{4^2}$$

$$F_{12}=k\frac{6q^2}{4} \qquad F_{32}=k\frac{q_3.2q}{16}$$

$$k\frac{6q^2}{4} = k\frac{q_3.2q}{16}$$

$$q_3 = 12q$$

Example: If the net force acting on q_2 is F_2 find the charge of q_3 in terms of q?

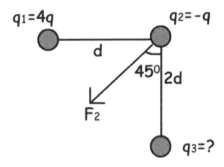

F_{12}
←――――┐
      │45°
      ↓
     F_2 F_{32}

Since angle is 45° then F_{12} must be equal to F_{32} in magnitude.

$F_{12} = F_{32}$

$$k\frac{q_1.q_2}{d_1^2} = k\frac{q_3.q_2}{d_2^2}$$

$$k\frac{4q.(-q)}{d^2} = k\frac{q_3.(-q)}{4d^2}$$

$$\Rightarrow q_3 = 16q$$

Example: If the system given below is in equilibrium, find q in terms of given quantities and tension in the rope in terms of mg.

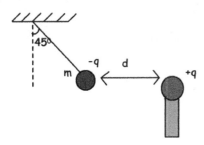

If the system is in equilibrium then net force acting on the –q must be zero.

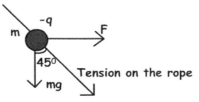

since the angle is 45°
F=mg

$$k\frac{q \cdot q}{d^2} = mg$$

$$q = d \cdot \sqrt{\frac{mg}{k}}$$

Tension on the rope can be found by using pisagor theorem

$T^2 = F^2 + (mg)^2$
$T = mg\sqrt{2}$

ELECTRIC FIELD

A charged particle exerts a force on particles around it. We can call the influence of this force on surroundings as electric field. It can be also stated as electrical force per charge. Electric field is represented with E and Newton per coulomb is the unit of it.

$$E = \frac{F}{q} = k\frac{q}{r^2}$$

Electric field is a vector quantity. And it decreases with the increasing distance. $k = 9 \cdot 10^9 \, N \cdot m^2/C^2$

- Electric field cannot be seen, but you can observe the effects of it on charged particles inside electric field.

• To find the electric field vector of a charge at one point, we assume that as if there is a +1 unit of charge there.

• If you want to find total electric field of more than one charges, you should find them one by one and add them using vector quantities.

Electric Field Lines

Motion path of "+" charge in an electric field is called electric field line. Intensity of lines shows intensity of the electric field. Pictures given below show the drawings of field line of the positive charge and negative charge.

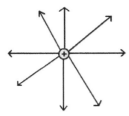 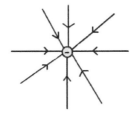

Electric field lines;

• Are perpendicular to the surfaces

• Never intercept

• If the electric field lines are parallel to each other, we call this regular electric field and it can be possible between two oppositely charged plates. E is constant within this plates and zero outside the plates.

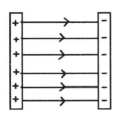

• We can find electric field between these plates by connecting a power supply having potential difference V by using following formula;

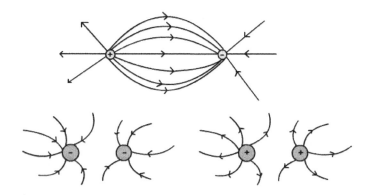

$$E\ (N/C) = \frac{V\ (volt)}{d\ (m)}$$

where, V is the potential diffrence of power supply,
d is the distance betwee the plates

Pictures given below show the path of the electric field lines of two same charges and two opposite charges.

Example: Find the electric field created by the charges A and B at point C in terms of $k.q/d^2$?

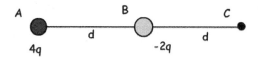

We assume that, there is a +q charge at C.

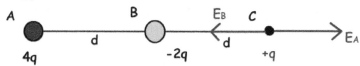

4q at point A repels the q at C and -2q at point B attracts the q at C, so directions of electric fields are opposite as shown in the picture.

$$E_A = k.\frac{4q}{4d^2} \qquad E_B = k.\frac{-2q}{d^2}$$

$$E_C = E_A + E_B = k.\frac{q}{d^2} - k.\frac{2q}{d^2}$$

$$E_C = -k.\frac{q}{d^2}$$

214

Example: If the electric field at point A is zero, find the charge at point D in terms of q.

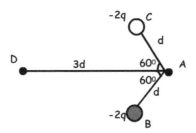

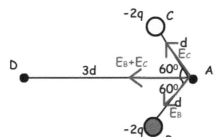

$E_B = k \cdot \dfrac{-2q}{d^2}$ $E_C = k \cdot \dfrac{-2q}{d^2}$

$E_B + E_C = k \cdot \dfrac{-2q}{d^2}$

Be careful! They are summed up by using vector quantities. As we learned in propoerties of vectors resultant vector of two equal vectors having 120° angle is equal to one of them.

Electric field of D must be equal to the sum of E_C and E_B in magnitude but opposite in direction. Thus;

$E_D = k \cdot \dfrac{2q}{d^2}$

$k \cdot \dfrac{q_D}{9d^2} = k \cdot \dfrac{2q}{d^2}$

$q_D = 18q$

FORCE ACTING ON A CHARGED PARTICLE INSIDE ELECTRIC FIELD

E=F/q

F=E.q where; F is the force acting on the charge inside the electric field E. Using this equation we can say that;

If **q** is positive then **F=+E.q** and directions of Force and Electric Field are same

If **q** is negative then **F=-E.q** and directions of Force and Electric Field are opposite

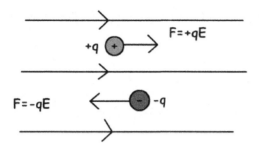

Example: If the charge q having mass m is in equilibrium between two plated having distance d, find the potential difference of power supply.

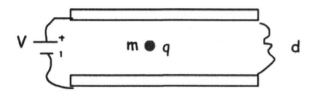

If charge is in equilibrium then force of gravity must be equal to the electric force.

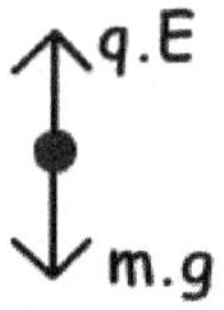

where; $E = V/d$

$$q\frac{V}{d} = m.g \qquad V = \frac{m.g.d}{q}$$

Electric Field of a Conductor Sphere

There is a maximum electric field at surface of the sphere. As distance increases from the surface, electric field decreases. Finally, as it seen from the picture, inside the conductor sphere electric field is zero.

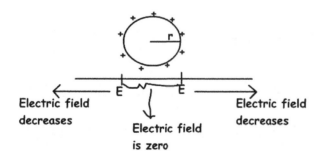

216

ELECTRIC POTENTIAL AND ELECTRIC POTENTIAL ENERGY

We learned that in work power energy chapter, objects have potential energy because of their positions. In this case charge in an electric field has also potential energy because of its positions. Since there is a force on the charge and it does work against to this force we can say that it must have energy for doing work. In other words, we can say that "Energy required increasing the distance between two charges to infinity or vice verse" is called electric potential energy. Electric potential energy is a scalar quantity and Joule is the unit of it. We use following formula to find the magnitude of EP;

$$\text{Electric Potential Energy} = k \frac{q_1 \cdot q_2}{d}$$

Be careful!

- In this formula if the charges have opposite sign then, Ep becomes negative; if they are same type of charge then, Ep becomes positive.

- If Ep is positive then, electric potential energy is inversely proportional to the distance d.

- If Ep is negative then, electric potential energy is directly proportional to the distance d.

Figure:1   Figure:2   Figure:3

In Figure 1 and Figure 2, charges repel each other, thus external forces does work for decreasing the distance between them. On the contrary, in Figure3, charges attract each other, electric forces decrease distance between them, and there is no need for other external forces.

Example: System given below is composed of the charges, 10q, 8q and -5q. Fin the total electric potential energy of the system.

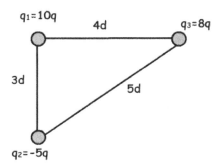

We should find the potential energies one by one taking two charges, and to find total energy we sum them up.

$$EP_{12}= k\frac{q_1 \cdot q_2}{d} = k\frac{10q \cdot (-5q)}{3d} = k\frac{-50q^2}{3d}$$

$$EP_{13}= k\frac{q_1 \cdot q_3}{d} = k\frac{10q \cdot 8q}{4d} = k\frac{20q^2}{d}$$

$$EP_{23}= k\frac{q_2 \cdot q_3}{d} = k\frac{(-5q) \cdot 8q}{5d} = k\frac{-8q^2}{d}$$

$$EP_{total}=EP_{12}+EP_{13}+EP_{23} = k\frac{-50q^2}{3d} + k\frac{20q^2}{d} + k\frac{-8q^2}{d}$$

$$EP_{total}=k\frac{-14q^2}{3d}$$

Electric Potential

Electric potential is the electric potential energy per unit charge. It is known as voltage in general, represented by V and has unit volt (joule/C).

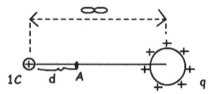

1C charge is brought to the point A from infinity. Work done here is called potential of q at A. Electric potential is found by the given formula;

V=k.q/d

V is a scalar quantity. If q is negative then V becomes negative, or if q is positive then V becomes positive.

Surfaces having equal potentials are called **equipotential surfaces**.

Potential of a Charged Sphere

Potential at surface is equal to the potential inside the sphere. Since there is no force acting inside the sphere, work is not done to bring the charge from surface to the inside of the sphere. As the distance from the surface of the sphere increase, potential decreases. Picture given below shows the change in the potential of the sphere inside, surface and outside. As you can see, potential is constant inside and surface of the sphere; however, it decreases with the increasing distance outside it.

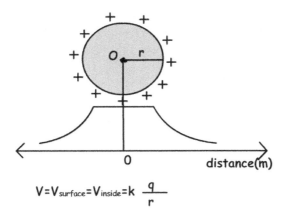

$$V = V_{surface} = V_{inside} = k\frac{q}{r}$$

Potential Difference between Two Points

Work done against to the electric field to move unit charge from one point to another is called potential difference between these two points. This difference is found by differences of potential of last point from initial point. If we take the point charge from A to B, then potential difference is found by the formula;

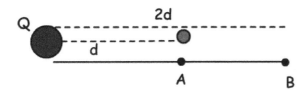

$$V_{AB} = V_B - V_A = \frac{W_{AB}}{q}$$

Example: Find the potential difference between points A and B, V_{AB} in terms of kq/r?

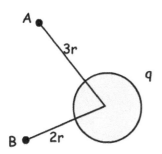

$$V_{AB} = V_B - V_A = \frac{W_{AB}}{q}$$

$$V_A = \frac{k.q}{3r} \qquad V_B = \frac{k.q}{2r}$$

$$V_{AB} = V_B - V_A = \frac{k.q}{r}\left(\frac{3}{6} - \frac{2}{6}\right)$$

$$V_{AB} = \frac{k.q}{6r}$$

Example: If the total electric field produced by q and q' is like in the picture given below, find the electric potential of the A.

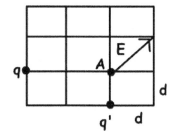

Since the degree of E with horizontal and vertical is 45°, magnitudes of Eq and Eq' must be equal.

$$E_q = k\frac{q}{(2d)^2} = k\frac{q}{4d^2} \qquad E_{q'} = k\frac{q'}{d^2} = k\frac{q}{d^2}$$

q'=4q and both q and q' are positive

$$V_A = V_1 + V' = \frac{k4q}{2d} + k\frac{q}{d} = 3k\frac{q}{d}$$

CAPACITANCE AND CAPACITORS

Capacitance can be defined as "gained charge per potential of conductors". Unit of capacitance is Coulomb per Volt and it is called as Farad (F).

$$C = \frac{Q}{V}$$

where; C is the capacitance of the conductor, Q is the amount of charge gained and V is the potential gained.

Capacitance is a scalar quantity. Graph given below shows the relation of a charged gained and potential gained of conductor sphere.

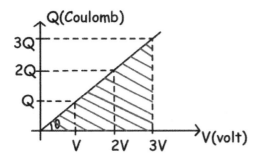

There is a linear relation between gained charge and gained potential. Slope of the graph gives us the capacitance of the sphere.

$$\tan\theta = \frac{Q}{V} = C$$

As I said before, farad is the unit of capacitance, however, we commonly use (pF) picofarad=10^{-12}F, (μF) microfarad=10^{-6}F and (nF) nanofarad=10^{-9}F.

Sphere having radius r and charge q has capacitance;

$$V = k\frac{q}{r}$$

$$C = \frac{Q}{V} = \frac{r}{q}$$

Capacitors

Capacitors are devices designed for storing charge. They are commonly used in computers or electronic systems. They consist of two conductor plates located with a distance to each other. When we connect negatively charged plate with neutral sphere, they share total charge until potentials become equal and leaves of the electroscope rise.

Then we locate plate A with a distance d to B. Since we ground the plate it is neutral at the beginning but since B is negatively charged it affects plate A and it is positively charged by induction.

If we put different insulator material between these plates like plastic leaves of electroscope are a little bit closed. We can conclude that, capacitance of the plates depends on the distance between them.

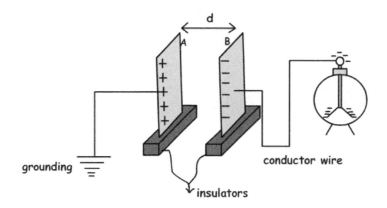

In a circuit we represent capacitor with the symbol;

Battery, which supplies potential difference, is represented by the symbol;

We show capacitors and battery in circuit as given below.

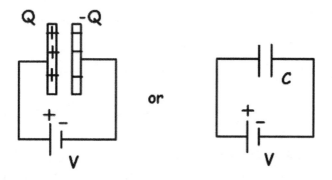

Capacitance of a plate depends on;

- Area of the plates

- Distance between the plates d

- Dielectric constant between the plates ε_0

As a result we can find capacitance of the plates with the following formula;

$$C = \varepsilon \frac{A}{d}$$

Dielectric constant between plates ε_0 depends on the type of material. For example, vacuum has $\varepsilon=8,85.10^{-12}$ F/m and water has $\varepsilon=717.10^{-12}$ F/m.

Example: Calculate the capacitance of the capacitor having dimensions, 30 cm X 40 cm and separated with a distance d=8mm air gap. (Dielectric constant of vacuum is $\varepsilon=8,85.10^{-12}$ F/m)

First we find area of the plate.

A=30.10⁻²mX40.10⁻²m=0,12m²

C= (8, 85.10⁻¹²C²/N.m²).0,12m²/8.10⁻³m

C=0, 13275.10⁻⁹ F

Capacitors in Series and Parallel

Systems including capacitors more than one has equivalent capacitance. Capacitors can be connected to each other in two ways. They can be connected in series and in parallel. We will see capacitors in parallel first.

In this circuit capacitors are connected in parallel.

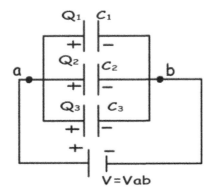

Because, left hand sides of the capacitors are connected to the potential a, and right hand sides of the capacitors are connected to the potential b. In other words we can say that each capacitor has same potential difference. We find the charge of each capacitor as;

$Q_1=C_1.V$

$Q_2=C_2.V$

$Q_3=C_3.V$

Total charge of the system is found by adding up each charge.

$Q_{total} = C_{eq} \cdot V$

$Q_{total} = Q_1 + Q_2 + Q_3 = C_1 \cdot V + C_2 \cdot V + C_3 \cdot V = V \cdot (C_1 + C_2 + C_3) = C_{eq}$

$C_{eq} = C_1 + C_2 + C_3$

As you can see, we found the equivalent capacitance of the system as $C_1 + C_2 + C_3$

Now we will see the capacitors in series.

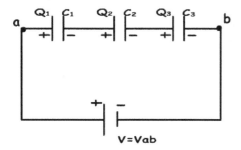

In capacitors in series, each capacitor has same charge flow from battery. In this circuit, +Q charge flows from the positive part of the battery to the left plate of the first capacitor and it attracts –Q charge on the right plate, with the same idea, -Q charge flows from the battery to the right plate of the third capacitor and it attracts +Q on the left plate. Other capacitors are also charged with same way. To sum up we can say that each capacitor has same charge with battery.

$C_1 \cdot V_1 = Q$

$C_2 \cdot V_2 = Q,$

$C_3 \cdot V_3 = Q$

$V = V_1 + V_2 + V_3$ and $Q = C_{eq} \cdot V$

$$\frac{Q}{C_{eq}} = \frac{Q}{C_1} + \frac{Q}{C_2} + \frac{Q}{C_3} = Q \left(\frac{1}{C_1} + \frac{1}{C_2} + \frac{1}{C_3} \right)$$

Equivalent Capacitance becomes;

$$\frac{1}{C_{eq}} = \frac{1}{C_1} + \frac{1}{C_2} + \frac{1}{C_3}$$

Example: Calculate the equivalent capacitance between points a and b.

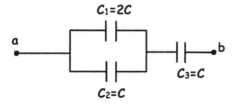

C_1 and C_2 are in parallel thus equivalence of them is;

$C_{12} = C_1 + C_2 = 2C + C = 3C$

C12 and C3 are in series thus equivalence of the is found by;

$$\frac{1}{C_{eq}} = \frac{1}{C_{12}} + \frac{1}{C_3} = \frac{1}{3C} + \frac{1}{C}$$

$$C_{eq} = \frac{3C}{4}$$

Example: In the circuit given below, C1=60μF, C2=20 μF, C3=9 μF and C4=12 μF. If the potential difference between points a an b Vab= 120V find the charge of the second capacitor.

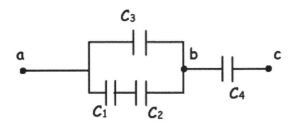

C_1 and C_2 are in series

$$\frac{1}{C_{12}} = \frac{1}{C_1} + \frac{1}{C_2}$$

$$\frac{1}{C_{12}} = \frac{1}{60} + \frac{1}{20} = \frac{4}{60} \qquad C_{12} = 15 \mu F$$

C_3 and C_{12} are in parallel

$C_{123} = C_{12} + C_3$

$C_{123} = C_{12} + C_3 = 15 + 9 = 24 \mu F$

C_{123} and C_4 are in series

$$\frac{1}{C_{eq}} = \frac{1}{C_{123}} + \frac{1}{C_4} = \frac{1}{24} + \frac{1}{12} = \frac{3}{24} \qquad C_{eq} = 8 \mu F$$

Total charge of the system is;
Q=C.V=8 μF.120V=960 μF

Potential difference between points B and C is;

$$V_{BC} = \frac{q}{C} = \frac{960}{12} = 80V$$

Potential difference between points A and B is;
$V_{AB} = V_{AC} - V_{BC} = 120V - 80V = 40V$
$Q_1 = Q_2 = C.V_{AB} = 15 \mu F . 40V = 600 \mu C$

MORE EXAMPLES RELATED TO ELECTROSTATICS

Example: If we touch two spheres to each other, find the final charges of the spheres.

Charge per unit radius is found;

$q_r=(Q_1+Q_2)/(r_1+r_2)$

$q_r=(20-5)q/(2r+r)=5q/r$

Charge of first sphere becomes;

$Q_1=q_r.r_1=5q/r.2r=10q$

Charge of second sphere becomes;

$Q_2=q_r.r_2=r.5q/r =5q$

Example: Positively charged sphere B is placed between two neutral spheres A and C. We cut connection of A and C with ground. If we put A closer to the first electroscope and touch C to the sphere of second electroscope, find the type of charge electroscopes have.

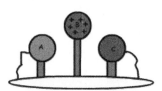
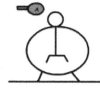

A and C are negatively charged by induction. Thus, leaves of both electroscopes are negatively charged.

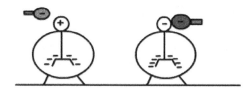

Example: If force applied by charge placed at point B on A is F, find forces applied by charges C and D on A in terms of F.

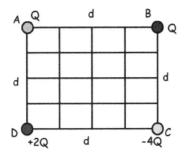

Free body diagram of forces is given below;

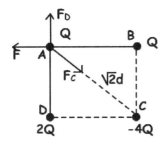

$F = k \cdot Q \cdot Q / d^2 = k \cdot Q^2 / d^2$

$F_C = k \cdot Q \cdot (-4Q)/(\sqrt{2}d)^2 = -4k \cdot Q^2/2d^2 = -2 \cdot k \cdot Q^2/d^2 = -2F$

$F_D = k \cdot Q \cdot 2Q/d^2 = 2k \cdot Q^2/d^2 = 2F$

Example: Find the electric field at point A produced by charges q_1 and q_2 in terms of k, q and d.

We assume that there is a +q charge at point A while finding electric field at point A.

$E_1 = k \cdot (-4q)/d^2 = -4k \cdot q/d^2$

$E_2 = k \cdot (16q)/4d^2 = 4k \cdot q/d^2$

Resultant electric field at point A is;

$E_{resultant} = E_1 + E_2 = -4kq/d^2 + 4kq/d^2 = 0$

Example: Find electric potential energy produced by Q_1, Q_2 and Q_3 in terms of $k.q^2/r$.

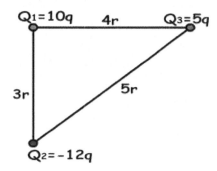

$E_P = k.Q_1.Q_2/r$

$E_{P1,2} = k.10q.(-12q)/3r = -40k.q^2/r$

$E_{P1,3} = k.10q.5q/4r = 25k.q^2/2r$

$E_{P2,3} = k.(-12q).5q/5r = -12k.q^2/r$

$E_P = E_{P1,2} + E_{P1,3} + E_{P2,3}$

$E_P = -40k.q^2/r + 25k.q^2/2r - 12k.q^2/r$

$E_P = -39{,}5k.q^2/r$

Example: Find the equivalent capacitance between points A and B.

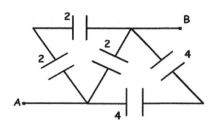

If we redraw circuit given above;

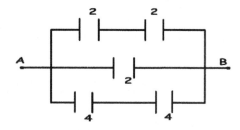

$1/C_{eq1} = 1/2C + 1/2C$

$C_{eq1} = 1C$

$1/C_{eq2} = 1/4C + 1/4C$

$C_{eq2} = 2C$

$C_{AB}=1C+2C+2C=5C$

Example: Find relation between the electrical energies stored in the capacitors.

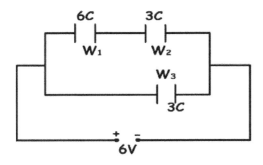

Since capacitors are connected in parallels, potentials in each branch are equal.

$W_1=1/2.6C.2^2=12C$

$W_2=1/2.3C.4^2=24C$

$W_3=1/2.3C.6^2=54C$

$W_3>W_2>W_1$

Example: Neutral sphere A, positively charged sphere B and negatively charged sphere C are given below. If we touch B inside of the A and C to outside of the A, find the final charges of spheres.

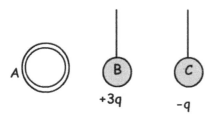

When we touch B inside of A, all charges of B are transferred to A and B becomes neutral. When we touch C outside of A, it is positively charged. Final charges of spheres;

A: is positively charged

B: is neutral

C: is positively charged

Example: A and B produce potential V at point X. If total potential at point X is -V, find q_C.

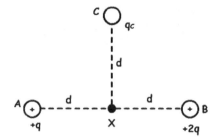

$V_{AB}=k.q/d+k.2q/d=3k.q/d=V$

$V_X=3k.q/d+k.q_C/d=-V$

$3k.q/d+k.q_C/d=-3k.q/d$

$q_C=-6q$

Example: If the capacitance of first capacitor is C, find the capacitance of second capacitor.

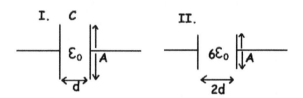

$C_1=\varepsilon_0.A/d=C$

$C_1=6\varepsilon_0.A/2/2d=6/4\varepsilon_0.A/d=3/2C$

Example: If we first touch A to B then A to C, find the final charges of these spheres. (Charges of spheres; 1st sphere: 8q, 2nd sphere: -2q, 3rd sphere: -14q)

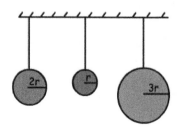

$q_A+q_B=8q-2q=6q$

Charge per unit radius;

$q_r=q_{total}/r_{total}=6q/(2r+r)=2q/r$

Charge of $q_{A'} = 2r \cdot 2q/r = 4q$

Charge of $q_B = r \cdot 2q/2 = 2q$

$q_{total} = q_{A'} + q_C = 4q - 14q = -10q$

$q_r = q_{total}/r_{total} = -10q/(3r+2r) = -2q/r$

Final charges of spheres:

$q_{Afinal} = 2r \cdot (-2q)/r = -4q$

$q_{Cfinal} = 3r \cdot (-2q)/r = -6q$

$q_B = r \cdot 2q/2 = 2q$

Example: Look at the picture given below. Find the tension in the rope.

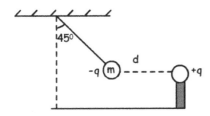

Free body diagram of the system is given below;

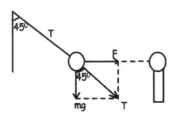

$\tan 45^0 = F/mg = 1$

$F = mg$

$k \cdot q \cdot q/d^2 = m \cdot g$

$q = d\sqrt{mg/k}$

We find T using Pythagorean theorem.

$T^2 = F^2 + mg^2$

$T^2 = (mg)^2 + (mg)^2$

$T = mg\sqrt{2}$

Example: A, B and C spheres are charged. If A repels B and attracts C, and B is positively charged, find the types of charges of each sphere.

If A repels B, then they must be charged with same charge; and if A attracts C then they must be charged with opposite charges.

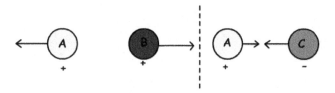

Example: We put negatively charged rod closer to neutral electroscope. If we ground electroscope, what will be the motion of the leaves of electroscope?

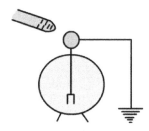

When we put negatively charged rod near the sphere of electroscope, sphere is positively charged by induction and negative charges are repelled to the leaves. When we ground electroscope, negative charges move to the ground and leaves becomes neutral and closed.

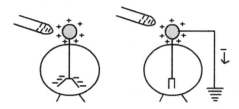

Example: Find charge of q_3 to make total force acting on q_2 zero.

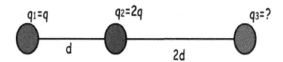

F_{12} must be equal F_{23}

k.q$_1$.q$_2$/d^2=k.q$_2$.q$_3$/(2d)^2

q$_1$/d^2=q$_3$/4d^2

q$_3$=4q$_1$=4q

ELECTRIC CURRENT

In the last unit we have learned static electric, or charge at rest. However, in this unit we will deal with the charges in motion. We see the electric current everywhere in daily life. Most of the electrical devices work with electric current. In this unit we will try to explain direction of the flow of current, ohm's law, and resistance of the electric circuit, resistors, measuring the current, and current density.

ELECTRIC BATTERY

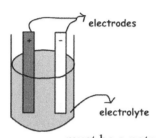

Electric battery is a device that converts chemical energy to the electrical energy. It consists of two different metal plates and we call them as electrodes. One of them is negatively charged *electrode* and other one is positively charged *cathode*. These plates are placed into a solution like dilute acid and it is also called electrolyte. This set up is called electric cell and connection of electric cells produce battery. There must be a potential difference between the electrodes to make charges flow.

ELECTRIC CURRENT AND FLOW OF CHARGE

If we connect a conductor to a battery, potential difference occurs between the ends of conductor. This potential difference creates an electric field towards to the positive end of the conductor to the negative end. Free charges inside this electric field are exerted a force F=q.E in this field. Under the effects of this force electrical charges starts to flow. This flow of charge is called **electric current**. If there is no potential difference then there won't be flow of charge or electric current. We can make an analogy with heat transfer. As we discussed in earlier chapters, heat flows from matter having higher temperature to the lower temperature. In this case charges flows from higher potential to the lower potential.

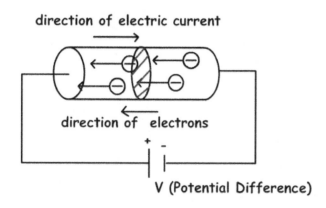

Electric current, in solids transferred with the free electrons, in liquids with free ions and in gases with free electrons and free ions. We can also define **electric current** as the charge per unit time passing through the cross section of conductor like given in the picture which is shown with red dashed lines. Average current **I** is found with the following formula;

$$I = \frac{\Delta Q}{\Delta t}$$

Where; I is the current, Q is the charge and t is the time

The unit of electric current is Coulomb per second, and we give specific name **Ampere (A)**.

Example: If the steady current 2,5 A flows in a wire for 5 minutes, find the charged passed in any point in the circuit.

$$\Delta Q = I \cdot \Delta t = (2,5 C/s)(300s) = 750C$$

OHM'S LAW RESISTANCE AND RESISTORS

Resistance is the difficulty applied by the conductor to the current flowing through it. Each material has different resistance. We show resistance with R and unit of it is ohm (Ω).

1 Ω = resistance of the conductor when 1 A current flows under the 1 V potential difference. Resistance is represented with the following picture in circuits;

—\/\/\/\—

Rheostat

Rheostat is a kind of device used to vary existing resistance. It is shown in the circuits as;

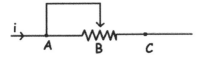

Resistance of the conductor depends on;

• Types of material or electrical resistivity of the material. It is shown with Greek letter ρ. Resistance of the material is linearly proportional to electrical resistivity.

• Length of the material (l). Resistance is linearly proportional to the length of the conductor.

- Cross sectional area of the conductor. Resistance is inversely proportional to the cross sectional area.

- Temperature. Temperature shows different effects with respect to the type of material.

We can write our formula with the explanation given above as;

$$R = \rho \frac{\ell}{s}$$

where; ρ is the electrical resistivity, ℓ is the length of the conductor and s is the cross sectional area of the conductor.

Ohm's Law

Ohm's law gives the relation between voltage, current and resistance. According to Ohm, current in a circuit is directly proportional to the applied voltage and inversely proportional to the resistance of the conductor. We can summarize this explanation with following formula;

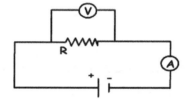

If we read the voltage values on voltmeter as V_1, V_2, V_3 and current on ammeter as I_1, I_2 and I_3, there is a relation between them as shown below;

$$\frac{V_1}{I_1} = \frac{V_1}{I_1} = \frac{V_1}{I_1} = \text{constant} = R$$

$$R = \frac{V}{I} \quad \text{or} \quad V = I.R$$

If we take resistance of the conductor constant, then potential difference and current of the system changes linearly as shown in the following graph;

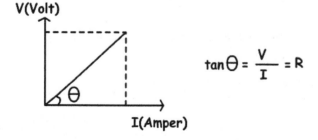

Example: Potential difference vs. Current graph of a conductor is given below, Find the behavior of resistance in intervals I, II and III.

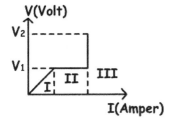

Ohm's law states that V=I.R

I. In the first interval, since the potential difference and current are linearly increases then resistance of the system becomes constant.

II. In the second interval, potential difference is constant however, current increases. This can be possible by decreasing in the amount of resistance.

III. In the third interval, current is constant, however, potential difference increases. This can be possible by increasing in the amount of resistance.

Combination of Resistors

Resistors can be combined in two ways; series and parallel. Combination of more than one resistor is called equivalent resistor. We first look at the resistors in series;

Resistors in Series

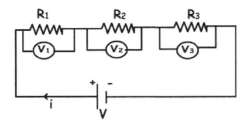

In these types of circuit, amount of currents passing through the resistors are equal and this current comes from the battery.

i=i$_1$=i$_2$=i$_3$

Sum of the potential differences of each resistor is equal to total potential difference of the circuit or potential difference between the ends of battery.

V=V$_1$+V$_2$+V$_3$

If,

$V_1 = i_1 \cdot R_1$

$V_2 = i_2 \cdot R_2$

$V_3 = i_3 \cdot R_3$

$V = V_1 + V_2 + V_3$

$i R_{eq} = i_1 \cdot R_1 = i_2 \cdot R_2 = i_3 \cdot R_3$

Since the currents are equal to each other;

$R_{eq} = R_1 + R_2 + R_3$

When you add new resistors to the circuit in series, equivalent resistance of the circuit increases.

Resistors in Parallel

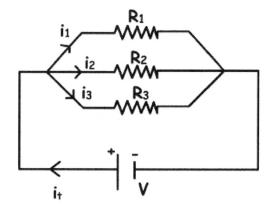

Sum of the currents in each branch is equal to the current coming from battery.

$i_t = i_1 + i_2 + i_3$

Since two ends of each resistor are connected to the same points, potential differences of each resistor are equal.

$V = V_1 = V_2 = V_3$

When you add new resistors to the circuit in parallel, equivalent resistance of the circuit decreases.

$$i_t = \frac{V}{R_{eq}} \quad i_1 = \frac{V_1}{R_1} \quad i_2 = \frac{V_2}{R_2} \quad i_3 = \frac{V_3}{R_3}$$

if we put these values into the following total current equation;

$i_t = i_1 + i_2 + i_3$

$$\frac{V}{R_{eq}} = \frac{V_1}{R_1} + \frac{V_1}{R_1} + \frac{V_3}{R_3}$$

Since the potential differences are equal;

$$\frac{1}{R_{eq}} = \frac{1}{R_1} + \frac{1}{R_1} + \frac{1}{R_3}$$

If n numbers of resistors are combined in parallel like in the picture given below, we find the equivalent resistance of the circuit as;

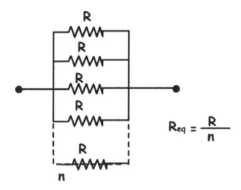

$$R_{eq} = \frac{R}{n}$$

Example: Find the equivalent resistance, if currents passing through each resistor and potential difference between the ends of each resistor of the circuit given below.

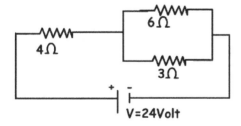

Since 3Ω and 6Ω resistors are in parallel, their equivalence becomes;

$$R_{eq1} = \frac{1}{6} + \frac{1}{3} = \frac{3}{6} \quad R_{eq1} = 2\,\Omega$$

Since 4Ω and R_{eq1} resistors are in series, their equivalence becomes;

239

$R_{eq} = 2\Omega + 4\Omega = 6\Omega$

We find current b using ohm's law
$V = i \cdot R_{eq}$
$24\text{volt} = i \cdot 6\Omega$
$i = 4\text{Amper}$

Since the equivalent resistance of 3Ω and 6Ω is 2Ω, potential difference between the ends **of this resistor is;**

$V = i \cdot R$

$V = 4A \cdot 2\Omega = 8\text{volt}$

$V = 4A \cdot 2\Omega = 8\text{volt}$
Currents in the resistors are;

$i_3 = 8\text{volt}/3\Omega = 8/3 \text{Amper}$

$i_6 = 8\text{volt}/6\Omega = 4/3 \text{Amper}$

$i_4 = 4\text{Amper}$

Potential difference between the ends of the 4Ω resistance is;

$V = i \cdot R$

$V = 4A \cdot 4\Omega = 16\text{volt}$

COMMON ELECTRIC CIRCUITS AND COMBINATION of BATTERIES

In common electric circuits you can come across with, batteries, switches, and connection wires. We now examine the components of the simple circuit one by one.

Batteries:
Batteries are devices, which supply energy to the circuit. We show it in the circuit as;

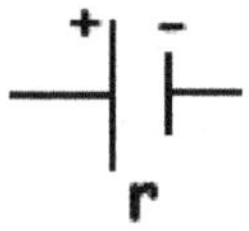

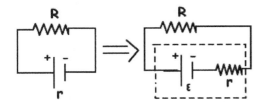

Where; **r** is the internal resistance of the battery and it behaves like as it is combined to the circuit in series and ε is the electromotive force (EMF) of the battery.

EMF:

It is the energy given by the battery to the unit charge when it passes from one end to the other end of the battery. If the battery gives W joule to a charge Q, then;

ε=W joule

We can say that, if the EMF of the closed circuit is known then EMF is directly proportional to the charge in the circuit.

W=ε.Q where Q=i.t

W=ε.i.t
Units are;

W=joule, ε=volt, i=Ampere and t=s.

Combination of Batteries

Batteries can be combined in three ways, in series, in parallels and in opposite directions.

Batteries in Series:

In this type of combination + end of the battery is connected to the – end of the other battery. Pictures given below show the examples of this type of combination.

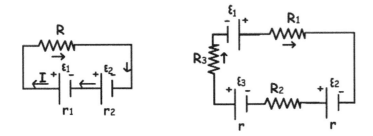

Total EMF of the circuit is;

$\varepsilon_{eq}=\varepsilon_1+\varepsilon_2+\varepsilon_3+...+\varepsilon_n$

Total internal resistance of the batteries is;

$r_{eq} = r_1 + r_2 + r_3 + ... + r_n$

Equivalent resistance of the circuit is;

Req=R+ r_{eq}

Batteries in opposite directions: is the combination of the batteries like shown in the figure below;

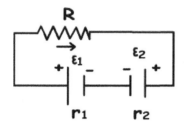

In this type of circuits, battery having larger EMF supplies the energy of the circuit and battery having smaller EMF behaves like a resistor. Total EMF of the circuit is found by;

If $\varepsilon_1 > \varepsilon_2$, $\varepsilon_{eq} = \varepsilon_1 - \varepsilon_2$

If $\varepsilon_1 = \varepsilon_2$, $\varepsilon_{eq} = 0$

If $\varepsilon_1 < \varepsilon_2$, $\varepsilon_{eq} = \varepsilon_2 - \varepsilon_1$

Total internal resistance of the batteries;

$r_t = r_1 + r_2$

Batteries in Parallel:

In this type of circuits, batteries must be identical. Picture given below shows the example of batteries in parallel;

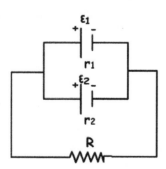

Total EMF of the circuit;

$\varepsilon_{eq} = \varepsilon_1 = \varepsilon_2 = \varepsilon$

Total internal resistance of the batteries is;

$$\frac{1}{r_{eq}} = \frac{1}{r_1} + \frac{1}{r_2} + \ldots + \frac{1}{r_n}$$

Example: In the circuit given below, there are identical batteries. Voltmeter reads 60V when the switch is opened and 50V when the switch is closed. Find the internal resistance of the batteries.

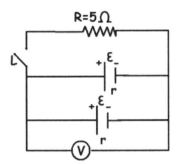

60 volt is the EMF of the circuit when switch is opened. When we close switch we read 50 volt thus;

50volt=5Ω.i

i=10 ampere

Potential difference between ends of the batteries, when the switch is closed, is;

50volt=60volt-10ampere.r/2 (r/2 is the equivalent resistance of the two batteries)

r=2Ω

Capacities of the batteries depend on the current passing through them. For example, look at the given circuits. In the first circuit, current i pass through the resistance and it passes through each battery, however, in the second circuit current i again pass through the resistance however, current i/2 pass through the each battery. Thus, you can use A' and B' batteries longer than A and B batteries.

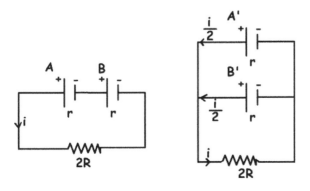

ALTERNATING CURRENT AND DIRECT CURRENT AND DIODES

DC means electrons flow only in one direction. Batteries, car batteries are examples of direct current. We analyze the structure of a battery in previous chapters. As you remember, battery has two terminal, "-"and "+". Electrons are attracted by the "+" terminal and flow of them starts in one direction. In DC, constant voltage supplied to the circuit since the flow is in one direction. Most of the portable devices using battery are work with direct current like CD player, calculator etc. This graph shows the change in the voltage with time. As you can see voltage of the circuit is constant.

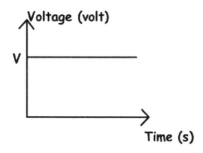

On the contrary, AC circuit means that, electrons do not flow in one direction. They move forward and back constantly. As a result of this changing motion of the electrons voltage of the system is also not constant. AC circuits show sinusoidal graphs as given below.

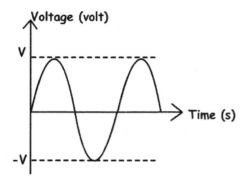

Most of the electrical devices in our homes work with alternating current, TV, washing machines, hair drier etc.

You can convert AC to DC with transformer, which decreases the amount of voltage, and you can use **diodes** in your circuits.

Diodes

Diodes are kinds of device that allow current flow only in one direction in circuits. Thus, only half of the cycles of alternating current can pass from the diodes. You can easily convert alternating current into the direct current with this device. Following circuit shows the usage of diodes;

In the given circuit, D1 let current flow however; D2 does not let current flow.

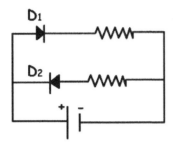

ELECTRIC POWER AND ENERGY

In a circuit given below, electrons coming from the battery transfer some of their energy to the conductor cable. They move and collide to the particles of conductor and this transferred energy converted into heat energy.

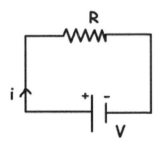

In a given unit of time we can find the emitted heat from the conductor as follows;

E=V.i.t

Where, **E** is the heat, **V** is the potential difference of the battery and t is the time. If we substitute i.R in terms of V then we find following energy equation.

E=i².R.t=V²/R.t

Unit of this energy is joule; if you want to convert it into calorie you can use the following equation;

1cal=4.18 joule

Electric Power

It is the energy emitted in a unit of time by the conductor.

Power=Electric Energy/time

If we substitute the energy formula, we get following equation for power.

P=V.i.t/t

P=i.V or if we put **i.R** in terms of **V** (ohm's law)

P=i².R=V²/R

Unit of the power is watt

1 watt=joule/s

Example: Find the relation between emitted energies by the resistors A, B, C and D in a unit of time.

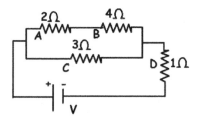

In this circuit, potential difference of the branches in parallel is equal. Currents passing through the branches are inversely proportional to the resistors. With these explanations, if the currents on the resistors A and B is i, then current on resistor C must be 2i. The sum of the currents 2i+i=3i coming from two branches pass through the last resistor D. Energy passing through a conductor in a unit of time is called power. Thus, powers of the resistors are;

From the equation **P=i².R**

$P_A = i^2 \cdot 2\Omega = 2i^2\Omega$

$P_B = i^2 \cdot 4\Omega = 4i^2\Omega$

$P_C = (2i)^2 \cdot 3\Omega = 12i^2\Omega$

$P_D = (3i)^2 \cdot 1\Omega = 9i^2\Omega$

Relation between them is like;

$P_C > P_D > P_B > P_A$

Example: A radiator working with 30V potential difference has power 180 watts. Find the current passing through and resistance of the radiator.

We use equation; **P=i.V**

180watt=i.30V

i=6 ampere

Resistance of the radiator **R=V/i=30V/6A=5Ω**

Example: If the current passing through this piece of circuit is i, power spent on the first resistance is 100watt. Find the potential difference between the ends of second resistor.

$$i \xrightarrow{\quad} \overset{R_1=4}{-\!\!\!\text{WWW}\!\!\!-} \overset{R_2=10}{-\!\!\!\text{WWW}\!\!\!-} \xrightarrow{\quad}$$

P₁=i².R₁

100watt=i².4Ω

i=5Ampere

Potential difference between the ends of second resistor is,

V=i.R

V=5Ampere.10Ω=50Volt

FINDING POTENTIAL DIFFERENCE BETWEEN TWO POINTS IN CIRCUITS

Potential difference between two points in circuit is the energy lost by the charge in being transferred from one point to another. For example, potential difference between A and B is found with following formula;

$$V_{AB} = V_B - V_A = \sum \varepsilon - i.R$$

This formula shows the energy lost by charge moving from point A to point B. We should first find the direction of the current to determine values. If you take the sign of batteries having same direction with current as "+" then you must take oppositely connected batteries as "-".

direction of the current →

$$\sum \varepsilon = -\varepsilon_1 - \varepsilon_2 + \varepsilon_3$$

If the direction of the current and current passing through the resistor are the same then we take i.R as "+", if they are in opposite directions then we take i.R as "-".

In the circuit given below, if the direction of the current is shown like below, then potential difference between the points A-B and C-B are;

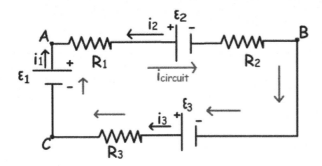

$V_{AB}=V_B-V_A= -\varepsilon_2-(+i.R_1+i.R_2)$

$V_{CB}=V_B-V_C= -\varepsilon_3-(-i.R_3)$

Example: Find the potential difference between points A and B in the given circuit below.

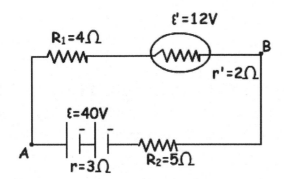

Current of the circuit is in the direction of battery current.

$i_{circuit}=\dfrac{\varepsilon-\varepsilon'}{R_1+R_2+r+r'} = \dfrac{40V-12V}{4\Omega+5\Omega+3\Omega+2\Omega}=2Amp$

$V_{AB}=-\varepsilon'-(i.R_1+i.r')=-12V-(4.2+2.2)=-12V-12V$
$V_{AB}=-24Volt$

MORE EXAMPLES RELATED TO ELECTRIC CURRENT

Example: Find the equivalent resistance of the circuit given below and current passing through the circuit.

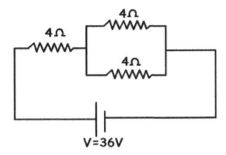

Since resistance are connected in parallel between points A and B;

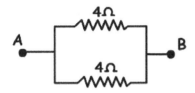

$1/R_{eq}=1/4+1/4$

$R_{eq}=2\Omega$

Equivalent resistance between points C and B;

C —4Ω—4Ω— B

$R_{eq}=4+2=6\Omega$

b) V=I.R

36=I.6

I=6Amperes

Example: Find equivalent resistance between the points X and Y and if the current passing from 7Ω resistor is I_1, and current passing from 8Ω resistor is I_2, find I_1/I_2.

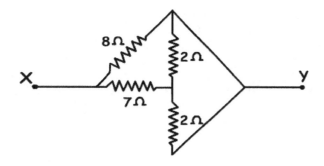

We redraw circuit and make it simple;

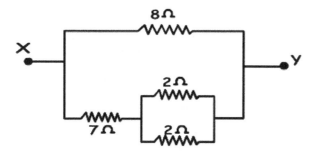

$R_{eq}=7+(2.2)/(2+2)$ (resistance in lower branch)

$R_{eq}=8\Omega$

Equivalent resistance between points X and Y;

$R_{eq}=(8.8)/(8+8)=4\Omega$

Since resistances are in parallel, their potentials are equal. Using ohm's law;

$V_1=I_1.R_1$ (lower branch)

$V=I_1.8$

$V_2=I_2.R_2$ (upper branch)

$V=I_2.8$

$I_1/I_2=1$

Example: If we close switch shown in the picture given below, find the changes in the brightness of the bulbs.

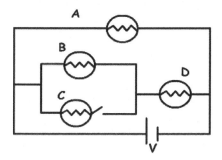

Circuit when switch is opened;

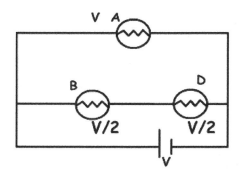

Potential of B is equal to potential of D.

$V_A > V_B = V_D$

When we close switch;

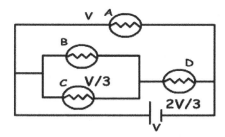

Potentials of the bulbs become;

$V_A = V$

$V_B = V_C = V/3$

$V_D = 2V/3$

Thus; there is no change in the brightness of the bulb A, brightness of B decreases, and D increases.

Example: Find the efficiency of the motor in the circuit given below.

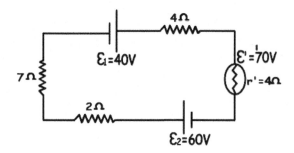

Current passing through circuit is;

I=(ε1+ε2-ε')/R_eq

I=(40+60-70)/(7+2+2+4)

I=2 Amperes

Efficiency=ε'/(ε'+I.r')=70/(70+2.2)

Efficiency=35/37

Example: Find the ratio of magnitude of two resistance made of same matter R_1/R_2?

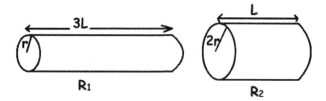

$R_1 = \rho \cdot 3L/(\pi \cdot r^2)$

$R_2 = \rho \cdot L/(\pi \cdot 4r^2)$

$R_1/R_2 = 12$

Example: Find the potential difference between points A and B.

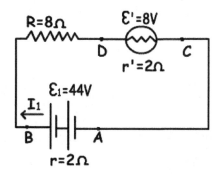

Current passing through circuit is;

I=(ε-ε')/(R+r+r')=(44-8)/(8+2+2)

I=3 Amperes

Potential between points D and C;

V$_{DC}$=-ε'-(Ir')=-8-(3.2)=-8-6=-14 Volt

Potential between points A and B;

V$_{AB}$=ε-(I.r)=44-(3.2)

V$_{AB}$=38 Volt

Example: Find potential difference between points X and Y.

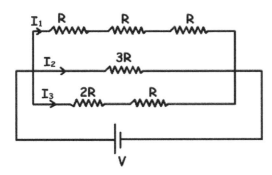

V$_{AB}$=V$_B$-V$_A$=Σε-ΣR.I

V$_{AB}$=V$_B$-V$_A$=ε-ε'-(I.R+I.r')

V$_{AB}$=50-30-2(6+4)

V$_{AB}$=20-20=0

Example: Find relation between heats produced by each branch of circuit given below.

Since branches are in parallel potential differences between them are equal.

I$_1$=V/3R, I$_2$=V/3R, I$_3$=V/3R,

I$_1$=I$_2$=I$_3$

$W = I^2 \cdot R \cdot t$

$W_1 = W_2 = W_3$ and heat produced;

$Q_1 = Q_2 = Q_3$

Example: If we close switches of circuits given below, find the changes in the brightness of the bulbs. (Neglect resistances of the batteries)

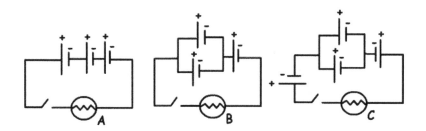

When we close switches, currents on A, B and C are;

$I_A = 3\varepsilon/R$

$I_B = 2\varepsilon/R$

$I_C = \varepsilon/R$

$I_A > I_B > I_C$

Brightness of the bulbs is directly proportional to currents passing on them.

$B_A > B_B > B_C$

MAGNETISM

In ancient times Greece people found a rock that attracts iron, nickel and cobalt. They call them as "magnet "and magnetism comes from here. Chinese people to make compasses used these rocks later. Later scientists found that, magnets have always two poles different from electricity. Magnets have two ends or faces called "poles" where the magnetic effect is highest. In last unit we saw that there is again two polarities in electricity, "-"charges and "+" charges. Electricity can exist as monopole but magnetism exists always in dipoles **North Pole (represented by N)** and **South Pole (represented by S)**. If you break the rock into pieces you get small magnets and each magnet also has two poles **N** and **S**.

Same poles of the magnet like in the electricity repel each other and opposite poles attract each other.

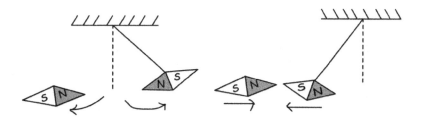

Strengths of these forces depend on the distance between the poles and intensity of the poles.

TYPES OF MAGNETS

In nature Fe_3O_4 is used as magnet. However, people can also produce magnets. They can have shapes rod, u shaped or horse shoe. Matters showing strong magnetic effect are called ferromagnetic; matters showing low magnetic effect are called diamagnetic and paramagnetic matters.

COULOMB's LAW FOR MAGNETISM

Effects of the two magnets to each other are inversely proportional to the square of distance between them and directly proportional to magnetic intensities of each magnet. These forces are equal in magnitudes and opposite in directions.

$$F = k \frac{m_1 \cdot m_2}{d^2}$$

Where; k is the constant, m_1 and m_2 are the magnetic intensities of the poles and d is the distance between them.

Example: Find forces exerted by the N poles of the magnets to each other. ($k=10^{-7} N.m^2/(Amp.m^2)$)

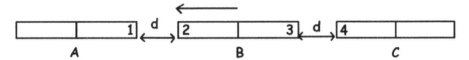

$$F = k\frac{m_1.m_2}{d^2} = \frac{10^{-7}.10^{-4}.10^{-3}}{0.5m} = 4.10^{-14}$$

Example: Three magnets are placed like given picture below. When the system released, magnet B gets closer to the magnet A. Find the possibilities of pole types of 1 and 4.

If we assume that 1 is **N** pole, then since 1 attracts 2, 2 must be **S**, 3 is **N** and 4 is **S**.

If we assume that 1 is **S** pole, then since 1 attracts 2, 2 must be **N**, 3 is **S** and 4 is **N**.

Example: Two magnets are placed like in the given picture below. If N pole of the magnet exerts F force on the N poles of the second magnet, find the net force exerted on the first magnet.

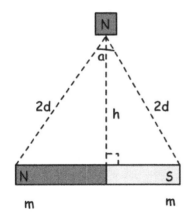

We use the similarities of triangles and get following equation.

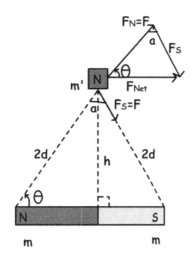

$$F_N = k\frac{m \cdot m'}{(2d)^2} = F \qquad k\frac{m \cdot m'}{d^2} = 4F$$

$$F_S = k\frac{m \cdot m'}{(2d)^2} = \frac{1.4F}{4} = F$$

$$\frac{F_{net}}{d} = \frac{F}{2d} \qquad F_{net} = \frac{F}{2}$$

MAGNETIC FIELD

Magnets show repulsion or attraction force around itself. This area affected from the force of magnets called **magnetic field**. We cannot see magnetic field necked eye. However, if we put a sheet on the magnet and put some iron filing on this sheet we can easily observe the magnetic field around the magnet with the shapes of the iron filling. The shapes of magnetic field lines showed in the picture given below;

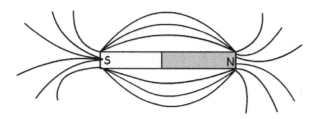

Directions of magnetic field lines are showed below;

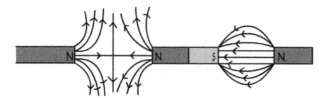

When the lines get closer to each other, this means that magnetic field is strong in that region.

Magnetic field lines;

- Never intersect

- If they are parallel we say that there is a regular magnetic field.

Electricity and magnetism are closely related to each other. Electricity causes magnetic field. Scientists have observed that, when a compass is put next to the wire (wire connected to a battery and current flows), needle of the compass is deflected. The reason of this deflection is magnetic field and they conclude that electric current produces magnetic field. Magnetic field lines around a wire are shown below;

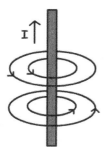

Magnetic field is a vector quantity and showed with the letter **B**. Unit of **B** is **Tesla**. When we calculate magnetic field of a magnet we assume that there is a 1 unit of m at the point we want to find. We find the magnetic field with following formula;

$$B = \frac{F}{m}$$

Where, F is the magnetic force and m is the magnetic intensities of poles.

Example: If k.m/d² is equal to 5 N/Amp.m find the magnetic field produced by the m_1 and m_2 at point A.

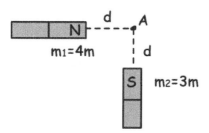

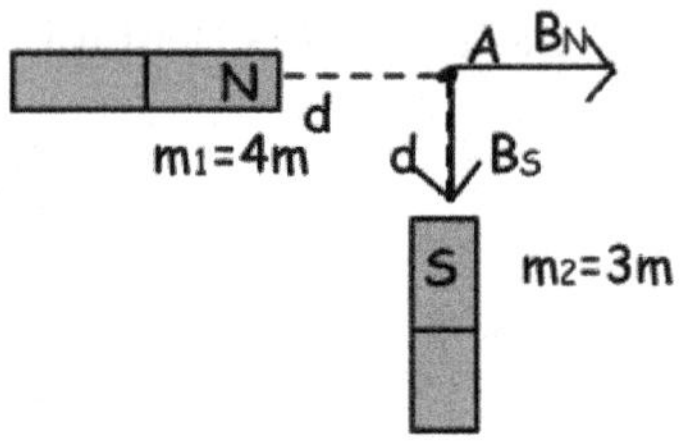

$B_N = k \dfrac{4m}{d^2} = 4.k\dfrac{m}{d^2} = 4.5 = 20 \dfrac{N}{Amp.m}$

$B_S = k \dfrac{3m}{d^2} = 3.k\dfrac{m}{d^2} = 3.5 = 15 \dfrac{N}{Amp.m}$

$B_A = B_N + B_S$ vector addition

$B_A = \sqrt{B_N^2 + B_S^2} = \sqrt{20^2 + 15^2} = \sqrt{400 + 225}$

$B_A = 25 \dfrac{N}{Amp.m}$

MAGNETIC FLUX

Magnetic flux is the number of magnetic field lines passing through a surface placed in a magnetic field.

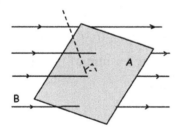

We show magnetic flux with the Greek letter; **Φ**. We find it with following formula;

Φ=B.A.cosΘ

Where Φ is the magnetic flux and unit of Φ is Weber (Wb)

B is magnetic field and unit of B is Tesla

A is area of the surface and unit of A is m^2

Following pictures show the two different angle situation of magnetic flux.

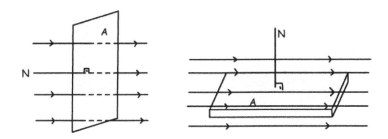

In the first one, magnetic field lines are perpendicular to the surface, thus, since angle between normal of the surface and magnetic field lines 0 and cos0=1 equation of magnetic flux becomes;

Φ=B.A

In the second picture, since the angle between the normal of the system and magnetic field lines is 90° and cos90°=0 equation of magnetic flux become;

Φ=B.A.cos90°=B.A.0=0

MAGNETIC PERMEABILITY

In previous units we have talked about heat conductivity and electric conductivity of matters. In this unit we learn magnetic permeability that is the quantity of ability to conduct magnetic flux. We show it with **μ**. Magnetic permeability is distinguishing property of the matter, and every matter has specific **μ**. Picture given below shows the behavior of magnetic field lines in vacuum and in two different matters having different **μ**.

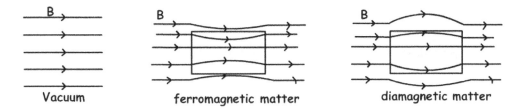

Magnetic permeability of the vacuum is denoted by; **μ₀** and has value;

$\mu_0 = 4\pi \cdot 10^{-7}$ Wb./Amps.m

We find the permeability of the matter by following formula;

μ= B / H where; H is the magnetic field strength and B is the flux density

Relative permeability is the ratio of a specific medium permeability to the permeability of vacuum.

$\mu_r = \mu/\mu_0$

Diamagnetic matters: If the relative permeability of the matter is a little bit lower than 1 then we say these matters are diamagnetic.

Paramagnetic matters: If the relative permeability of the matter is a little bit higher than 1 then we say these matters are paramagnetic.

Ferromagnetic matters: If the relative permeability of the matter is higher than 1 with respect to paramagnetic matters then we say these matters are ferromagnetic matters.

MAGNETIC FIELD OF EARTH

A bar magnet or a compass when hanged from their gravitational center, they come equilibrium at north-south pole direction of the earth. This situation shows that there must be magnetic field acting on compass and bar magnet. This magnetic field is the earth's magnetic field. The reason for this magnetic field is not explained yet however, most of the scientists believe that, earth's magnetic field looks like magnetic field of a bar magnet. Picture given below shows the magnetic field lines of the earth;

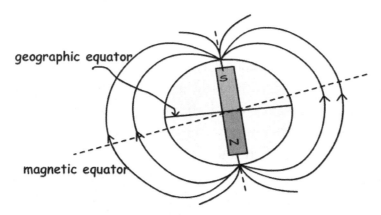

The geographic poles and magnetic poles do not coincide actually. There is an 11° angle between the geographic poles and magnetic poles. We call magnetic declination this deviation. Picture given below shows the geographic poles and magnetic poles of earth.

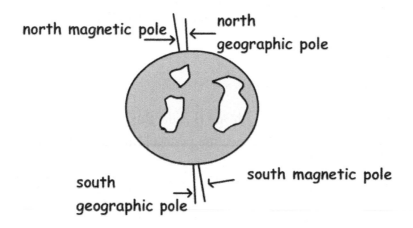

MAGNETIC EFFECT OF THE CURRENT

In 1800s scientist Hans Oersted observe that, current flowing in a circuit effects the direction of needles of the compass. Picture, given below, shows his experiment. When the switch is closed current passes the circuit and direction of the magnet changes under the effect of magnetic field produced by current.

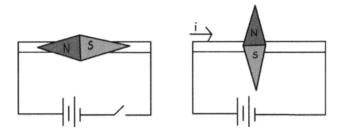

Moreover, Faraday and Joseph Henry are other scientists showing the relation of magnetic field and current. If you move the magnet placed near the circuit you produce current or, if you change the current of circuit you can get current in another circuit placed near it. We will learn all the types of current produced by magnetic field.

Magnetic Field around a Wire

Current flowing in a linear wire produces magnetic field **B=2k.i/d** at a distance d. Here current measured in Ampere, distance measured in meter and $k=10^{-7}$ N/Amps². Direction of the magnetic field produced around the wire is always tangential to the circle around the wire.

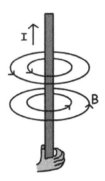

We can find the direction of magnetic field by using right hand rule. As shown in the picture above, grab the wire with your four fingers, your thumb shows the direction of current and four fingers show the direction of magnetic field. We show current in two ways, if the current towards to us we show it with a dot, if the current is outward we show it with cross.

inward   upward

Example: Two currents flow through the x and y axis of the wire. As you can see from the picture two points are located near the wires A and B. If the total magnetic field at **A** is **B$_A$**, and total magnetic field at **B** is **B$_B$**, find ratio of **B$_A$/B$_B$**?

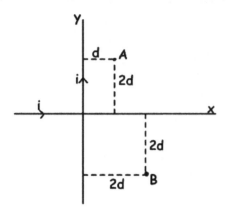

Magnetic field at point B;

$$B_{PX} = 2k \cdot \frac{i}{2d} \odot \quad \text{and} \quad B_{PY} = 2k \cdot \frac{i}{d} \otimes$$

Total magnetic field at P;

$$B_{PY} - B_{PX} = 2k \cdot \frac{i}{2d} \otimes$$

Magnetic field at point R;

$$B_{RX} = 2k \cdot \frac{i}{2d} \otimes \quad \text{and} \quad B_{RY} = 2k \cdot \frac{i}{2d} \otimes$$

Total magnetic field at R;

$$B_R = 2k \cdot \frac{i}{d} \otimes$$

$$\frac{B_P}{B_R} = \frac{1}{2}$$

Example: As shown in the figure given below, i_1 current produces 8N/Amps.m magnetic field at point K. Find the magnitude and direction of the total magnetic field produced by i_1 and i_2 at point L.

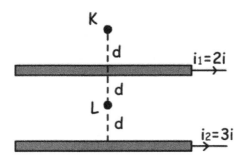

Directions of the currents are shown in the picture below;

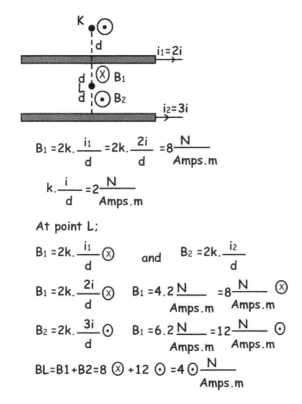

$B_1 = 2k \cdot \dfrac{i_1}{d} = 2k \cdot \dfrac{2i}{d} = 8 \dfrac{N}{Amps.m}$

$k \cdot \dfrac{i}{d} = 2 \dfrac{N}{Amps.m}$

At point L;

$B_1 = 2k \cdot \dfrac{i_1}{d}$ ⊗   and $B_2 = 2k \cdot \dfrac{i_2}{d}$

$B_1 = 2k \cdot \dfrac{2i}{d}$ ⊗ $B_1 = 4 \cdot 2 \dfrac{N}{Amps.m} = 8 \dfrac{N}{Amps.m}$ ⊗

$B_2 = 2k \cdot \dfrac{3i}{d}$ ⊙ $B_1 = 6 \cdot 2 \dfrac{N}{Amps.m} = 12 \dfrac{N}{Amps.m}$ ⊙

$B_L = B_1 + B_2 = 8$ ⊗ $+ 12$ ⊙ $= 4$ ⊙ $\dfrac{N}{Amps.m}$

Magnetic Field around a Circular Wire

Circular wire produces magnetic field inside the circle and outside the circle. Magnetic field around a circular wire is calculated by the formula;

B=2πk.i/r

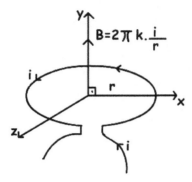

Direction of the magnetic field at the center of the circle is found with right hand rule.

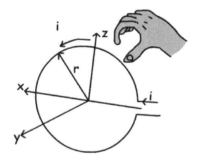

Your thumb shows the direction of magnetic field and four fingers show direction of current. Moreover, we can show the direction of current inside the circle with following pictures;

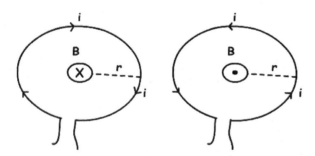

Example: Find the magnitude and direction of magnetic field at the center of the semicircle given below.

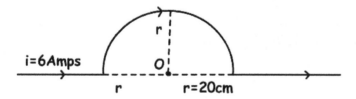

When we apply right hand rule we see that direction of magnetic field is inward to the page as shown in the picture below, since we have semicircle, we put 1/2 in front of our formula;

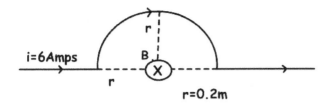

$B = \frac{1}{2} 2\pi k \frac{i}{r}$

$B = \frac{1}{2} 2 \cdot 3 \cdot 10^{-7} \frac{6}{0.2}$

$B = 9 \cdot 10^{-6} \frac{N}{Amps \cdot m}$

Example: Directions of i_1 and i_2 currents are opposite. If the magnetic field at the center of the circles is zero find the ratio of i_1 to i_2 i_1/i_2?

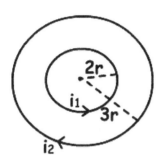

Smaller circle has magnetic field;

$B_1 = 2\pi k \frac{i_1}{2r}$ ⊙

magnetic field of larger circle;

$B_2 = 2\pi k \frac{i_2}{3r}$ ⊗

Since;

$B_1 + B_2 = 0$ magnitudes of magnetic fields are equal to each other;

$B_1 = B_2$

Thus;

$\frac{i_1}{i_2} = \frac{2}{3}$

Example: Find the magnetic field produced by currents i_1 and i_2 at point O.

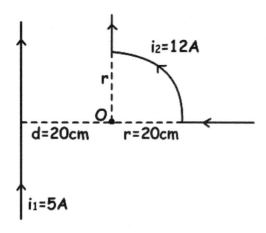

If we apply right hand rule, directions of currents are;

i_1: inward

i_2: outward

Thus, total magnetic field at point O becomes the difference of these magnetic fields.

$$B_1 = 2k\frac{i_1}{d} = 2 \cdot 10^{-7} \frac{5}{0.2} = 5 \cdot 10^{-6} \frac{N}{Amps.m} \quad \otimes$$

$$B_2 = \frac{1}{4} 2k\pi \frac{i_2}{r} = \frac{1}{2} \cdot 3 \cdot 10^{-7} \cdot \frac{12}{0.2} = 9 \cdot 10^{-6} \frac{N}{Amps.m} \quad \odot$$

$$B = B_1 + B_2$$

$$B = B_2 - B_1 = 4 \cdot 10^{-6} \frac{N}{Amps.m} \quad \odot$$

Magnetic Field around a Solenoid

Picture given below shows the solenoid. A typical solenoid behaves like a bar magnet. Magnetic field produced by solenoid is constant inside the solenoid and parallel to the axis of it.

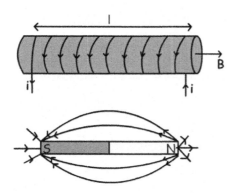

We find the magnetic field produced by solenoid with the following formula;

$$B = 4\pi k \frac{i \cdot N}{l}$$

Where: i is the current, N is the number of loops and l is the length of the solenoid.

We find the direction of magnetic field by using right hand rule again. Grab the solenoid as your four fingers show the direction of current and your thumb shows the direction of magnetic field.

Example: Find the magnetic field produced by the solenoid if the number of loop is 400 and current passing through on it is 5 A. (length of the solenoid is 40 cm k=10^{-7}N/Amps²)

$$B = 4\pi k \frac{i \cdot N}{l} = 4.3 \frac{10^{-7} \cdot 5 \cdot 400}{0.4}$$

$$B = 6 \cdot 10^{-3} \frac{N}{Amps \cdot m}$$

Example: A solenoid has 80 cm diameter, number of loop is 4 and magnetic field inside it is 1,2·10^{-5}N/Amps.m. Find the current passing through each loop of wire.

$$i_{loop} = n \cdot i \qquad r = \frac{80}{2} = 40cm = 0.4m \qquad B = 1,2 \cdot 10^{-4} \frac{N}{Amps \cdot m} \qquad n = 4$$

$$B = 2\pi k \frac{i}{r}$$

$$B = 2 \cdot 3 \cdot 10^{-7} \cdot \frac{4 \cdot i}{0.4}$$

$$i = 2 Amps$$

Example: There are two solenoid given below, they have equal lengths and i_1=4Amps and i_2=3Amps. Find the magnetic field vector at point A.

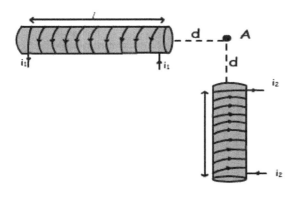

i_1 current produces B_1 magnetic field and i_2 current produces B_2 magnetic field. We sum these vectors using vector properties and get following total magnetic field vector at point A.

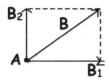

FORCE ACTING ON MOVING PARTICLE AND CURRENT CARRYING WIRE

As we learned before, charged particles produce electric field around themselves. In an electric field charged particles are exerted force F=qE. The motion of the charges in an electric field produce current and as a result of the current magnetic field is produced. This magnetic field exerts force on the charged particles inside the field. Experiments done on this subject show that we can find the force exerted on the current carrying wire with following formula;

F=B.i.l.sinß

Where B is the magnetic field strength, i is the current and l is the length of the wire and ß is the angle between magnetic field and the wire.

We find direction of the force by right hand rule. Picture given below shows the direction of magnetic field current and force;

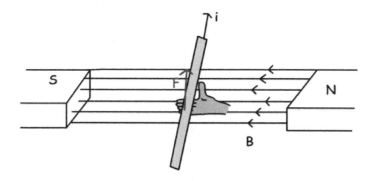

If the angle between current and magnetic field is **ß**;

1. **ß=0 then sinß=0, F=0**

2. **ß=180 then sinß=0, F=0**

3. **ß=90 then sinß=1, F=B.i.l**

We can say that, if the directions of current and magnetic field are parallel to each other then, no force exerted on the wire.

Example: Which one of the magnetic force acting on the wires is/are zero given in the picture below?

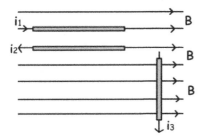

Since the directions of the currents i_1 and i_2 are parallel to the direction of magnetic field, no force exerted on these currents.

F₁=F₂=0

i_3 current is perpendicular to the magnetic field thus,

F₃=B.i₃.l

Direction of the magnetic force is toward us.

Example: Find the directions of the magnetic forces acting on the currents i1, and i2 placed in a constant magnetic field.

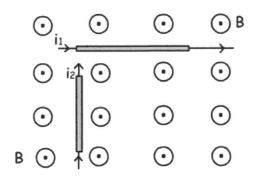

Magnetic forces acting on the currents i1 and i2 are shown in the picture below.

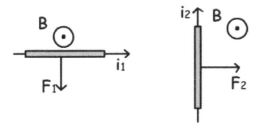

FORCE ACTING ON CHARGED PARTICLE

Force acting on a current is explained above. We have learned that the motion of charged particles produces current. Thus, force on current carrying wire is the sum of forces acting on each charged particle, which produce this current. If the particle has charge q, velocity v and it is placed in a magnetic field having strength B force acting on this particle is found with following formula;

F=q.v.B

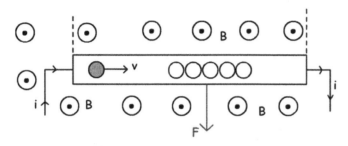

If;

1. v=0, then F=0 no force exerted on stationary particle in magnetic field.

2. ß=0, then sin0=0 and F=0

3. ß=180, then sin180=0, and F=0, magnetic field lines and velocity of particle parallel to each other, then no force exerted on it.

4. ß=90, then sin90=1, F=q.v.B

FORCES OF CURRENT CARRYING WIRES ON EACH OTHER

Experiments done on this subject shows that currents in the same direction attract each other since they produce opposite magnetic fields. On the contrary currents in opposite directions repel each other since they produce magnetic fields having same directions.

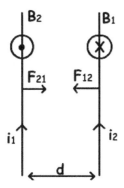

We find force exerted on each of them with following formula;

$$F = B_1 \cdot i_2 \cdot l = 2 \cdot k \frac{i_1 \cdot i_2}{d} l$$

Where; l is the length of the wires, d is the distance between them.

TRANSFORMERS

Transformers are devices used for changing the potential of alternating currents. Structure of simple transformer is given below;

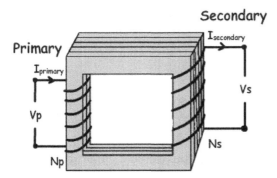

Voltage is applied to primary windings and, we take transformed voltage from secondary windings. There are two types of transformer, step up and step down. We use step down transformers in electrical devices like radio, and step up transformers in welding machine.

Step Up Transformer

This type of transformer used for increase the incident voltage. Number of turns in secondary windings is larger than the number of turns in primary windings.

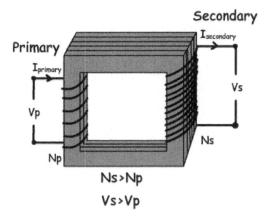

Step Down Transformer

This type of transformer used for decrease incident voltage. Number of turns in primary windings is larger than the number of turns in secondary windings.

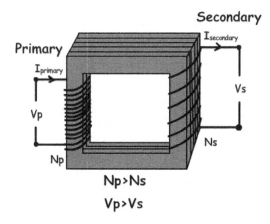

Transformer Equations

Vp is the potential, Ip is the current, Np is the turn on the primary windings and Vs is the potential, Is is the current, Ns is the turn on the secondary windings. We use following equations to find potential, current or number of turns of any windings;

$N_1/N_2 = V_1/V_2 = I_2/I_1$

MORE EXAMPLES RELATED TO MAGNETISM

Example: Find the forces exerted by S poles of magnets given below.

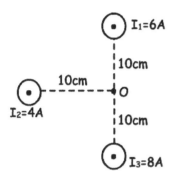

$F = k \cdot M_1 \cdot M_2 / r^2 = (10^{-7} \cdot 10^{-4} \cdot 10^{-3})/(0{,}6)^2$

$F = 10^{-14}/(36 \cdot 10^{-2})$

$F = 10^{-12}/36$

Example: Find resultant magnetic field at point O, produced by I_1, I_2 and I_3.

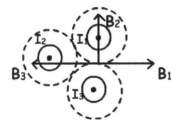

Magnitudes of magnetic fields;

$B_1 = 2k \cdot 6/0{,}1 = 12 \cdot 10^{-7}/10^{-1} = 12 \cdot 10^{-6}$ N/Amps.m

$B_2 = 2k \cdot 4/0{,}1 = 8 \cdot 10^{-7}/10^{-1} = 8 \cdot 10^{-6}$ N/Amps.m

$B_3 = 2k \cdot 8/0{,}1 = 16 \cdot 10^{-7}/10^{-1} = 16 \cdot 10^{-6}$ N/Amps.m

$B_{resultant} = B_1 + B_2 + B_3$

$B_{resultant} = \sqrt{(12 \cdot 10^{-6} - 16 \cdot 10^{-6})^2 + (8 \cdot 10^{-6})^2}$

$B_{resultant} = 4\sqrt{5} \cdot 10^{-6}$ N/Amps.m

Example: A, B and C wires are given below. Find the magnetic field of A, B and C at points X and Y.

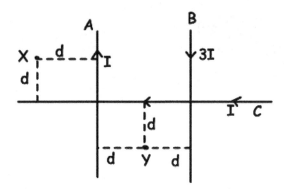

Directions of magnetic fields at point X are found using right hand rule.

B_A: outward

B_B: inward

B_C: inward

$B_X = B_B + B_C - B_A$

$B_X = 2k \cdot 3I/3d + 2k \cdot I/d - 2k \cdot I/d = 2k \cdot I/d$

Directions of magnetic fields at point Y are;

B_A: inward

B_B: inward

B_C: outward

$B_Y = B_A + B_B - B_C$

$B_Y = 2k \cdot I/d + 2k \cdot 3I/d - 2k \cdot I/d = 2k \cdot 3I/d$

Ratio of magnetic fields;

$B_X/B_Y = 2k \cdot I/d / 2k \cdot 3I/d = 1/3$

Example: Solenoid having number of loops N and surface area A is shown in picture given below. If we change the position of solenoid as shown in the picture below, find the equation used for finding induced emf of solenoid.

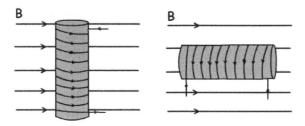

Induced emf=ε=-(ΔΦ)/(Δt).N

Change in Flux;

ΔΦ=Φ₂-Φ₁

$Φ_1=0$, since cross section area of solenoid and magnetic field lines are parallel to each other.

Φ₂=B.A

ΔΦ=B.A-0=B.A

ε=-B.A.N/t

Example: Draw the directions of magnetic field lines at point A, B, C and D in the picture given below.

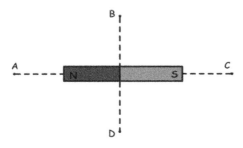

Directions of magnetic field lines are drawn from N pole to S pole as shown in the picture given below.

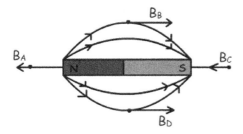

Example: Look at transformer given below and find whether the transformer is step down or step up, V_1/V_2 and whether secondary voltage depends on R or not.

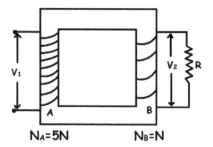

Since the number of turns in primary side is larger than the secondary side, transformer is step down.

$V_1/V_2 = N_A/N_B = 5N/N = 5$

$V_1/V_2 = N_A/N_B$

$V_1 = V_2 \cdot N_A/N_B$

Thus, V_1 does not depend on resistor R.

Example: If the system given below is in equilibrium, find the poles of magnets.

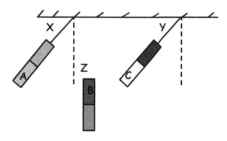

B attracts C and repels A. Thus;

If B is N, then C must be S and A must be N

If B is S, then C must be N and A must be S

Example: Force F acting on the AC wire, placed in a magnetic field B, is shown in the picture. Find the force acting on CB wire having length L.

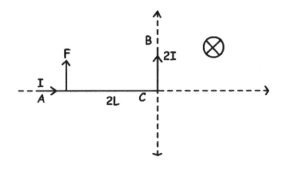

We find direction of force by right hand rule.

Direction of force is outward.

F_{AC}=B.I.2L=F

F_{CB}=B.2I.L=F

So, force acting on CB wire has magnitude F and, it is in -X direction.

Example: Three wires A, B and C are given below. If force acting on the one part of B exerted by A is F. Find resultant forces acting on wire C.

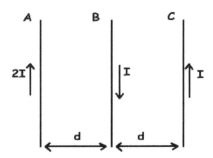

Since directions of currents on A and B are opposite in directions, A repels B.

F_{AB}=2k.(2I.I/d).L=F

A attracts C, B repels C

F_{AC}=2k.(2I.I/2d).L

F_{BC}=2k.(I.I/d).L

F_{AC}=F_{BC}

Resultant force acting on C is zero.

Example: If the magnetic field at point O is zero, find the ratio of r_1/r_2.

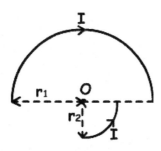

Magnetic field of circle having radius r_1;

$B_1 = 1/2 \cdot (2\pi \cdot k \cdot I / r_1)$ inward

Magnetic field of circle having radius r_2;

$B_2 = 1/4 \cdot (2\pi \cdot k \cdot I / r_2)$ outward

Since resultant magnetic field at point O is zero;

$B_1 = B_2$

$1/2 \cdot (2\pi \cdot k \cdot I / r_1) = 1/4 \cdot (2\pi \cdot k \cdot I / r_2)$

$r_1 / r_2 = 2$

WAVES

In this unit we will discuss properties of waves and types of waves. Moreover, we will try to explain situations, which cannot be explained with light properties of matter. Disturbance of the shape of the elastic matters are transported from one end to other by the particles of that matter, we call this process as wave. Be careful during this transportation no matter is transported.

Waves are classified in different ways with their properties. For example, mechanical waves and electromagnetic waves are classified according to the medium. Water waves and sound waves are examples of mechanical wave on the contrary, light waves; radio waves are examples of electromagnetic waves. Electromagnetic waves can propagate in vacuum.

Waves can propagate in 1D, 2D and 3D. Spring waves are examples of 1D wave, water waves are examples of 2D waves and light and sound waves are examples of 3D waves.

We can examine ways according to their propagation direction under two title; longitudinal waves and transverse waves.

Transverse Wave:

In these types of waves, directions of wave and motion of particles are perpendicular to each other. Picture given below shows this wave type.

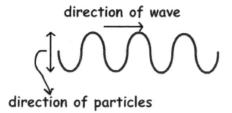

Longitudinal Wave:

In this type of waves, direction of the particles and wave are same. Look at the given picture below.

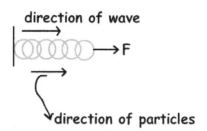

PROPERTIES OF WAVES

In this section we will learn basic properties of waves. We have learned transverse and longitudinal waves in last section. Now we use them and try to explain basic concepts of wave phenomena as; wavelength, velocity, amplitude, pulse, frequency.

Amplitude and Pulse: It is one wave motion created at the spring.

Where; x is the pulse length and y is the amplitude (height of the pulse).

Wavelength: It is the distance between two points of two waves having same characteristics.

Wavelength is shown with the Greek letter "λ" and unit of it is "m".

Period: Time required for production of one wave is called period. It is shown with letter "T" and its unit is "s".

Frequency: It is the number of waves produced in a given unit of time. It is shown with letter "f" and its unit is "1/s".

Periodic Wave: If the wave source produces equal number of waves in equal times, then this wave called periodic wave.

$f = 1/T$

Velocity of the Wave: Velocity of the wave is constant in a given medium. However, if the medium is changed then velocity of the wave is also changed. We show velocity with "v" and its unit is m/s.

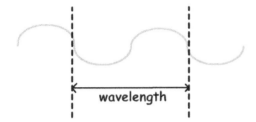

$$v = \frac{x}{t} = \frac{Distance}{Time} = \frac{\lambda}{T} = \lambda \cdot f$$

DIRECTION OF WAVE PROPAGATION

If we know shape of the pulse at an instant time, or propagation direction of the particles of the pulse we can find the direction of wave propagation.

Example: Given picture below shows the direction of wave propagation. Find the directions of the vibration at points A, B and C.

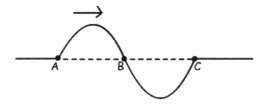

We draw the shape of the pulse after t s and find the directions of the vibration at points A, B and C.

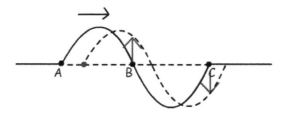

VELOCITY OF THE SPRING PULSE

Distance taken at an instant time by the pulse is called velocity of the pulse. Velocity of the spring pulse depends on force exerted on the spring and spring constant μ. Spring constant depends on the type of the spring; it is found by the following formula;

μ=mass/length

Where; v velocity of the spring pulse (m/s),

μ is the spring constant (kg/m), and F is the force exerted on the spring (N).

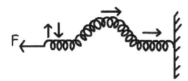

Example: Find the relation between propagation velocities of pulses of identical springs in given picture below.

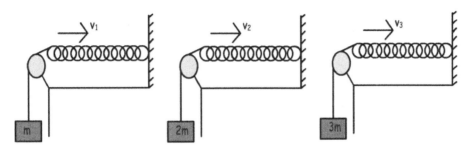

Forces exerted on the springs are directly proportional to the hanged masses. Thus, since $G_3 > G_2 > G_1$,

$v_3 > v_2 > v_1$.

Example: There are three identical springs having equal masses and different lengths $L_3 > L_2 > L_1$. Find the relation of velocities of pulses.

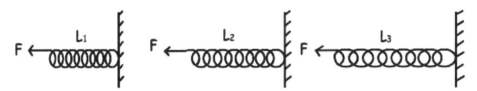

Propagation velocity of the pulse is;

$$v = \sqrt{\frac{F}{\mu}}$$

Spring constant is;

μ=mass/length

Relation of the lengths,

L₃>L₂>L₁

Relation of the spring constants;

μ₃<μ₂<μ₁

Propagation velocities become;

$v_3 > v_2 > v_1$

VELOCITY OF PERIODIC WAVES

We have learned that, if the medium is constant than velocity of the wave is also constant and we gave following equations for velocity of waves;

$$v = \frac{x}{t} = \frac{\text{Distance}}{\text{Time}} = \frac{\lambda}{T} = \lambda \cdot f$$

Example: Find the relation of wavelengths of given waves.

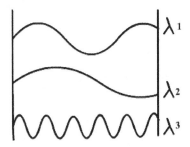

We define wavelength as the distance between the sequential crests or troughs. Picture given below shows the wavelengths of each wave and relation between them.

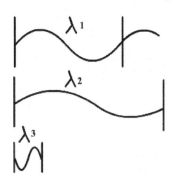

Example: Using the data given in the picture below; find the wavelength, velocity and amplitude of the wave. Frequency of the source is $2s^{-1}$.

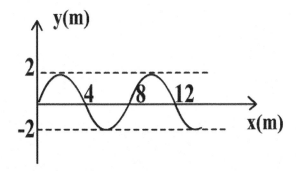

We find wavelength of the wave from the picture as;

8m

We find velocity of the wave by using the following formula;

v=wavelength.frequency=8m.2s^{-1}=16m/s

Amplitude of the wave is 2m from the given picture.

INTERFERENCE OF SPRING WAVES

When two waves interfere they produce resultant wave. In this section we learn how to find resultant wave. Displacement of the resultant wave is the sum of the waves producing it. Look at the given pictures below. They show the behavior of the waves before the interference, and after the interference.

Example: Amplitudes of the waves given below are A_1 and A_2. Picture shows the interference of two waves.

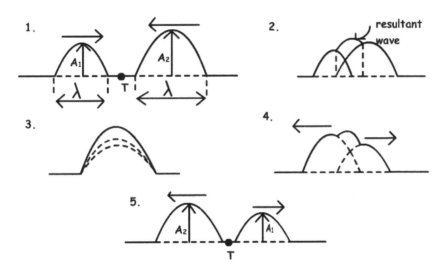

Picture given below shows two identical waves interference having opposite directions.

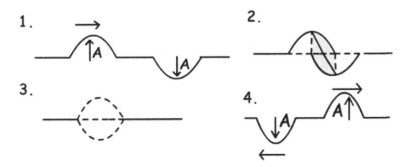

 Be Careful!

If the amplitudes of the waves having opposite directions are different, then amplitude of the resultant wave becomes the difference of the amplitudes of waves. It has the same direction with the bigger wave.

REFLECTION OF SPRING WAVES

Hitting an obstacle and turning back of the wave is called reflection of wave. We examine reflection of sprig waves under two title; reflection from a fixed end and reflection from an opened end.

Reflection from Fixed End

When a pulse of spring wave hits an obstacle having fixed end, it reflects. Reflected wave has opposite direction, same amplitude and velocity with the incident wave. Picture given below shows this process.

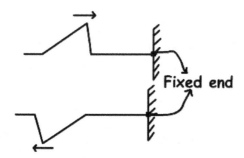

Reflection from Open End

When a pulse of a spring comes to an open-end obstacle it reflects, like given picture below. Amplitude, velocity and length of the pulse do not change but its right hand side becomes left hand side.

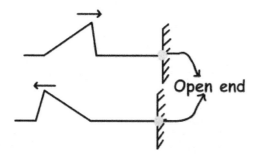

Reflection from High Density to Low Density and Low Density to High Density

We add two springs having different thicknesses and send a pulse from the spring having low density to high density. Some part of the pulse is transferred to the high-density spring and continues its motion and rest of the pulse reflects. Joining point of two springs behaves like fixed end obstacle. Picture given below shows the behavior of incident, transferred and reflected wave.

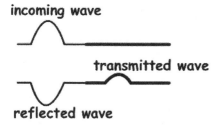

Relation between velocities of incident transmitted and reflected wave;

$V_{incident} = V_{reflected} > V_{transmitted}$

When a pulse send from the high-density spring to low-density spring, some part of the pulse reflects again and some part of it is transmitted. Picture given below shows the behavior of reflected and transmitted pulse.

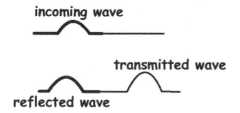

Amplitudes of the incident wave and reflected wave are equal but amplitude of transmitted wave is larger than them. Relation between the velocities of incident reflected and transmitted waves;

$V_{incident} = V_{reflected} < V_{transmitted}$

Example: Draw the transmitted and reflected wave of the given pulse below.

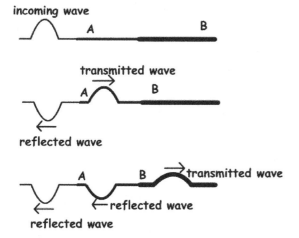

Example: Find the directions of the pulses A and B if the directions of the particles x and y given below.

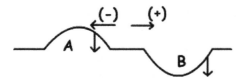

Given picture below shows the pulse A, and its shape after t second. From the positions of points x and y, we can say that this pulse travels along (-) direction and pulse B travels along (+) direction.

Example: There are two waves having equal length and amplitude like in the given picture below. Find the shape of the waves when they overlap.

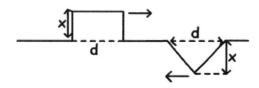

When two pulse overlap their shape becomes;

WATER WAVES

Glass water tanks are used for examining water waves and its properties. With the help of refraction of light, properties of water waves will be explained. You can produce two types of water waves, circular and linear. Picture, given below shows how we use light and determine wavelength of the water waves.

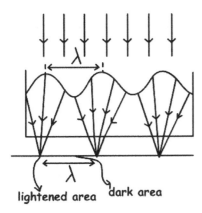

Crests of the waves behave like converging lens and we see lightened area at the bottom of the tank. On the contrary, troughs make us see darken area at the bottom of the water tank.

We can create linear waves using a rod, and circular waves using a point source.

Reflection of Linear Water Waves

Linear waves reflect from a linear surface with an angle equal to incident angle. Pictures given below show the incident wave and reflected wave with their angles.

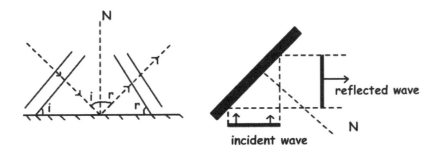

Here linear waves reflect from a circular surface like in the given picture below. After reflection their shape becomes the reflection surface's shape.

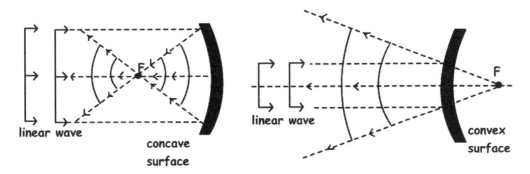

In the first picture, waves converge at one point and they turn into circular waves like in the concave mirrors. However, in the second picture, waves reflect from the circular surface as if they are coming from a point behind the surface like in the convex mirrors. Linear waves in the second picture also turn into circular wave.

Reflection of circular wave can be explained like reflection of light from curved mirrors. Pictures, given below, show some of the reflection of circular waves.

I. Picture given below shows the reflection of circular wave from concave surface. In this picture, waves come from the center of the curvature.

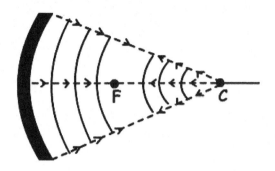

II. Picture given above shows the reflection of circular wave from concave surface. Waves come from the focal point of the surface and reflected circular waves become linear waves.

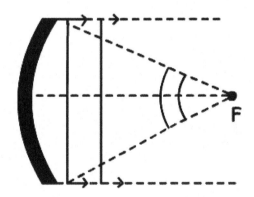

REFRACTION OF WAVES

Waves change direction when passing from one medium to another. This change in the direction of wave is called refraction of wave. During refraction velocity and wavelength of waves change however, frequency of waves stay constant. Velocity and wavelength of wave, coming from deep part of water tank to shallow part, decrease. Picture given below shows this change in the velocity and wavelength of waves.

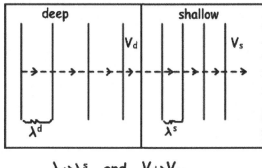

$\lambda_d > \lambda_s$ and $V_d > V_s$

If direction of waves coming from deep part of water tank is not perpendicular to the normal of the surface, then directions of refracted waves change. Look at the given picture below, it shows incident wave, refracted wave and angles between them.

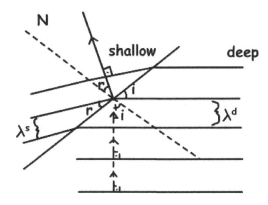

We can write following equations for incident and refracted waves;

$$\frac{\sin i}{\sin r} = \frac{V_d}{V_s} = \frac{\lambda_d}{\lambda_s}$$

Example: Velocity of the wave in the deep part of the tank is 16m/s. Picture given below shows the refraction of this wave when it pass to the shallow part of the tank. Find the wavelength of this wave in shallow part of the tank.

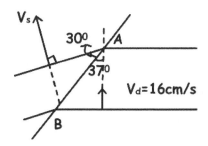

293

$$\frac{\sin 53°}{\sin 30°} = \frac{V_d}{V_s} \qquad \frac{0,8}{0,6} = \frac{16}{V_s}$$

$V_s = 10 cm/s$

$f_{source} = f_d = f_s = 2s^{-1}$

$\lambda_s = V_s/f = 10/2 = 5 cm$

Example: If the side view of the water tank is given in the picture below; draw the top view of the periodic linear waves produced by the source.

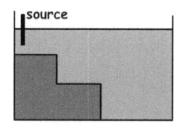

There are three different depths in this tank. Thus, velocity and wavelength of the deep part is bigger than the velocity and wavelength of the shallow part.

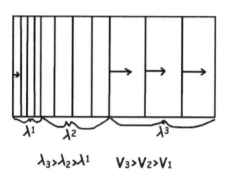

$\lambda_3 > \lambda_2 > \lambda_1 \qquad V_3 > V_2 > V_1$

Example: If the top view of the water tank is given below, find the shape of the linear wave after refracting from the deep part of the tank.

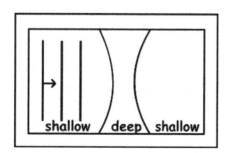

First two ends of the linear wave enter to the deep water, so velocity of these parts of wave increase and linear wave becomes circular wave.

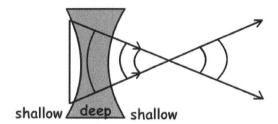

MORE EXAMPLES RELATED TO WAVES

Example: Pulses given below move distance X in one second. Find when they interfere?

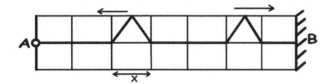

Pulse in left hand side of the water tank reflects from open end obstacle and pulse in right hand side of the water tank reflects from fixed end obstacle. After 5,5 seconds later they interfere.

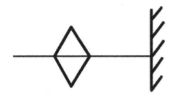

Example: Two different spring having different densities are added at point B. If the pulse send from one of the spring is reflected and transmitted from point B, find whether the following statements are true or false. ($L_1 < L_2$)

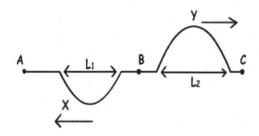

I. Y is reflected pulse

II. Density of AB spring is smaller than density of BC spring

III. Incoming wave comes from AB.

Velocity of the pulse in denser spring is smaller than the velocity of pulse in lighter spring. Since, $L_1 < L_2$, AB is the denser spring so, II is false. Since shapes of pulses are different we can say that some part of incoming wave is reflected and some part of it is transmitted. I is true and III is false

Example: If distance between 5 crests is 60 cm and frequency of the wave source is 3s⁻¹, find velocity of wave.

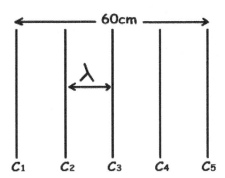

4λ=distance between 5crests

4λ=60cm

λ=15cm

λ.f=V

15.3=V

V=45cm/s

Example: If the waves A and B coming from point O, to the curved surface, they reflect as shown in the picture given below. Find whether the following statements are true or false.

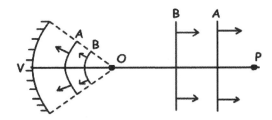

I. O is center of the curved surface

II. O is focal point of the curved surface

III. P is the center of the curved surface

Since waves coming from point O, reflects from the curved surface and move parallel, point O becomes focal point of the curved surface. I is false and II is true.

We cannot say anything about the center of the curved surface, thus III may be true or false.

Example: Which ones of the following statements change wavelength of the wave.

I. Changing frequency of wave source

II. Moving wave source

III. Changing water level of tank

I, II and III change wavelength of the wave.

Example: Picture given below shows wave motion of source having frequency $2s^{-1}$.

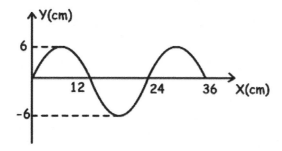

Find wavelength, velocity and amplitude of wave.

Using picture given above, we find wavelength as;

24cm

Since;

λ.f=V

24.2=V

V=48 cm/s

Using picture given above, we find amplitude as;
A=6 cm

Example: Springs having different thicknesses are attached at point A. If the pulse having wavelength λ_2 is transmitted pulse, find the relation between V_1, V_2, λ_1 and λ_2.

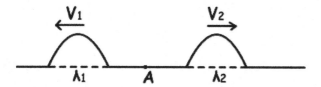

Since both of the pulses have same shape, pulse comes from spring having thicker one to thinner one. Pulse having velocity V_2 is transmitted wave, thus $V_1<V_2$.

$V_1 = \lambda_1 \cdot f$

$V_2 = \lambda_2 \cdot f$

$\lambda_1 < \lambda_2$

Example: If F is the focal point of curved obstacle, draw the reflected waves coming from wave source A.

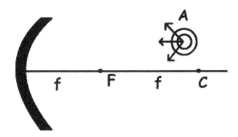

Curved obstacle behaves like concave mirror. Image of incoming waves are located away from the center of the obstacle.

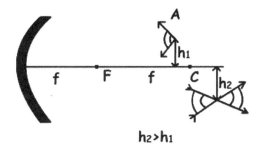

$h_2 > h_1$

Example: Top view of water tank is given below. Draw the shapes of waves shown in the picture below in deep part of water tank.

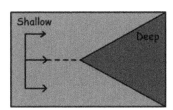

Waves in deep part of water tank has larger velocity thus, shape of waves becomes;

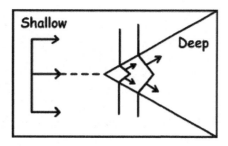

Example: Point A on wave becomes crest of wave after 3 second. Find period of wave.

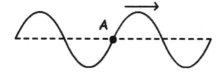

Period of the wave;

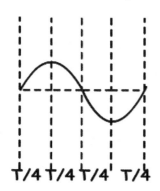

3T/4=3s

T=4s

INDEX

acceleration, 8, 18, 22, 23, 24, 27, 28, 29, 30, 31, 32, 33, 34, 35, 36, 40, 41, 42, 43, 54, 55, 62, 63, 64, 65, 68, 69, 71, 72, 73, 74, 75, 92, 94, 100, 103, 114, 117, 118, 119, 122

amplitude, 283, 286, 288, 289, 290, 298

angular speed, 114, 116

apparent depth, 155, 156

area expansion, 189, 190, 192

Batteries, 240, 241, 242, 244

Capacitance, 220, 222

Capacitors, 221, 223

centrifugal force, 121, 122

collisions, 101, 107, 111

concave mirrors, 136, 137, 147, 159, 160, 291

Conduction, 180, 202

conductors, 181, 201, 204, 220

conservation of energy, 95

contraction, 190

Convection, 181

convex mirror, 135, 144, 146, 160, 161

critical angle, 152, 153, 154

current, 4, 38, 234, 235, 236, 237, 238, 243, 244, 245, 246, 247, 248, 249, 259, 263, 264, 266, 269, 270, 271, 272, 274

density, 150, 167, 168, 169, 170, 172, 173, 174, 234, 261, 288, 289, 296

diamagnetic, 256, 262

diodes, 244, 245

displacement, 8, 18, 19, 20, 21, 28, 29, 30, 31, 34, 57, 74, 75, 78, 116

distance, 18, 19, 20, 21, 32, 35, 36, 38, 39, 42, 44, 47, 48, 49, 50, 51, 55, 56, 78, 79, 80, 83, 84, 85, 89, 92, 93, 99, 114, 115, 122, 123, 133, 134, 135, 136, 142, 144, 146, 155, 159, 160, 161, 164, 183, 184, 189, 205, 207, 209, 210, 212, 216, 217, 218, 221, 223, 256, 257, 263, 273, 283, 286, 296, 297

elasticity, 167, 171, 175

Electric battery, 234

Electric field, 212, 213

electric potential, 217, 218, 220, 228

Electric Power, 246

electrical force, 212

Electrons, 200, 204, 209, 244

electroscope, 205, 206, 207, 208, 209, 221, 226, 232

energy, 4, 82, 83, 84, 85, 86, 87, 88, 89, 90, 93, 94, 95, 96, 107, 110, 177, 179, 180, 181, 182, 183, 189, 217, 218, 228, 234, 240, 241, 242, 245, 246, 247

equilibrium, 59, 60, 61, 69, 96, 212, 216, 262, 278

Evaporation, 185

ferromagnetic, 256, 262

focal point, 136, 137, 138, 139, 140, 143, 144, 145, 147, 160, 161, 162, 292, 297, 299

force, 8, 16, 53, 54, 55, 58, 59, 60, 61, 62, 63, 64, 65, 66, 67, 68, 69, 71, 72, 73, 74, 75, 76, 78, 79, 80, 81, 84, 85, 91, 93, 94, 100, 101, 102, 103, 104, 105, 109, 111, 118, 119, 121, 122, 123, 128, 167, 171, 185, 209, 210, 211, 212, 215, 217, 218, 226, 232, 234, 241, 257, 258, 259, 270, 271, 272, 273, 278, 279, 284

free fall, 18, 32, 33, 34, 35, 43, 44, 45, 46, 53, 54, 55, 90, 95

freezing, 178, 183, 184, 194

frequency, 114, 115, 117, 126, 129, 283, 286, 287, 292, 297, 298

friction, 67, 68, 69, 72, 73, 86, 87, 89, 91, 93, 96, 119, 120, 122, 202

grounding, 204, 205

Heat, 179, 180, 181, 182, 183, 194, 197

heterogeneous mixtures, 169

homogeneous mixtures, 169, 174

Impulse, 100, 101, 109

Impurity, 185, 187

Inertia, 122, 167

instantaneous speed
  instantaneous speed, 21

insulators, 201

linear expansion, 189, 190, 191

magnetic field, 258, 259, 260, 261, 262, 263, 264, 265, 266, 267, 268, 269, 270, 271, 272, 275, 276, 277, 278, 279, 280

magnetic flux, 260, 261

Magnetic permeability, 261

magnetism, 4, 256, 259

mass, 8, 62, 63, 64, 83, 85, 89, 91, 93, 96, 99, 100, 101, 102, 104, 105, 106, 107, 108, 109, 111, 115, 116, 118, 119, 121, 129, 167, 168, 169, 170, 173, 174, 175, 179, 180, 182, 184, 185, 186, 188, 190, 194, 196, 216, 284, 285

melting, 175, 177, 183, 184, 185, 186, 188, 189

momentum, 4, 99, 100, 101, 102, 103, 104, 105, 107, 108, 110, 114

Newton, 53, 58, 62, 63, 64, 68, 69, 72, 103, 118, 119, 122, 212

normal force, 65, 66, 67

Ohm's law, 236, 237

Optic, 131

paramagnetic, 256, 262

period, 36, 100, 114, 115, 116, 119, 126, 127, 283, 300

plane mirrors, 131, 133, 137

potential energy, 82, 83, 84, 85, 87, 88, 95, 189, 217

power, 4, 81, 213, 216, 217, 246, 247

principal axis, 136, 137, 138, 139, 143, 144, 145, 146, 159, 160, 161

projectile motion, 38, 46, 53, 54, 55, 57

Radiation, 181

real depth, **155**, **156**
Reflection, 131, 132, 288, 291, 292
refraction, 131, 147, 148, 150, 151, 152, 155, 290, 292, 293
relative motion, 18, 36, 38, 105
Resistance, 105, 235, 236, 247
Rheostat, **235**
Riverboat problems, **38**
scalar, **8**, **9**, **10**, 18, **19**, 22, 63, 64, 78, 99, 217, 218, 220
solenoid, 268, 269, 276, 277
specific heat capacity, 179, 180, 182, 183
speed, **18**, **19**, **20**, **21**, 22, 24, 32, 33, 53, 58, 99, 114, 115, 116, 117, 118, 119, 121, 131, 135, 148, 149, 181, 183, 189
sublimate, **186**
tangential speed, **114**, 115, 118
temperature, 4, 8, 168, 175, 177, 179, 180, 181, 182, 183, 184, 185, 186, 187, 188, 189, 190, 191, 192, 193, 195, 196, 197, 198, 234
torque, 114, 122, 123, 124
total reflection, **152**, 153, 154
transformer, **244**, 273, 274, 278

vector, **8**, **9**, **10**, **11**, **12**, **13**, **14**, **15**, 18, 20, 22, 24, 36, 38, 48, 53, 58, 60, 63, 64, 99, 100, 105, 107, 118, 122, 212, 213, 259, 269, 270
velocity, 8, 18, **20**, 21, 22, 23, 24, 25, 26, 27, 28, 29, 30, 31, 32, 33, 34, 35, 36, 37, 38, 39, 40, 41, 42, 43, 44, 45, 46, 47, 48, 49, 50, 51, 54, 55, 57, 58, 62, 75, 85, 86, 87, 88, 89, 93, 94, 95, 96, 99, 100, 101, 102, 104, 105, 106, 107, 108, 109, 110, 111, 114, 115, 116, 117, 118, 120, 122, 126, 127, 128, 134, 135, 149, 150, 158, 159, 272, 283, 284, 285, 286, 287, 288, 292, 294, 295, 296, 297, 298, 299
vertex, 136, 137, 140, 143, 145, 160
virtual image, 133, 140
Volume, 167, 169, 174, 175, 193
volume expansion, 189, 190, 193
water waves, 282, 290
wave, 131, 181, 282, 283, 284, 286, 287, 288, 289, 291, 292, 293, 294, 295, 296, 297, 298, 299, 300
Wavelength, 283
work, 3, 4, 78, 79, 80, 81, 82, 83, 85, 86, 91, 92, 93, 181, 217, 218, 234, 244

Made in the USA
Columbia, SC
27 April 2021